THE TOXIC CLOUD

THE
TOXIC
CLOUD

Michael H. Brown

1817

HARPER & ROW, PUBLISHERS, New York

Cambridge, Philadelphia, San Francisco, Washington
London, Mexico City, São Paulo, Singapore, Sydney

FIRST EDITION

Designer: C. Linda Dingler
Copy editor: Rick Hermann
Indexer: Sylvia Farrington

Library of Congress Cataloging-in-Publication Data

Brown, Michael Harold, 1952–
   The toxic cloud.

   Includes index.
   1. Air—Pollution—Environmental aspects—United States. 2. Pollutants—Environmental aspects—United States. 3. Environmental chemistry—United States. 4. Hazardous wastes—Environmental aspects—United States. I. Title.
TD883.2.B66   1987   363.7'392'0973   87-45027
ISBN 0-06-015801-8

---

87 88 89 90 91   RRD 10 9 8 7 6 5 4 3 2 1

For my grandparents
Also, for those who have fought for their lives,
and for those who have lost them

CONTENTS

"Before destruction the heart of man is haughty . . ."

—*Proverbs 18:12*

PART I

GHOSTS OF THE MIDWEST

1

The winds come from the west, sluggish and foggy when the Pacific is at peace, the air up high moving steadily, relentlessly eastward, the lower draft butting against sea cliffs, drifting slowly over the mussel coves.

In day the mass of air is an off-white soup of sulfides and hydrocarbons, infused, over the metropolises, with filaments of russet brown.

At sunrise, in cities like Los Angeles, the hue is a resplendent purple that eventually makes its way east over the desertlands of Nevada and Arizona, settling as a haze in the Grand Canyon and reaching even into Utah (when the weather is right) before funneling through mountain passes or lofting into the turbulence above.

There the winds eddy and dip or wash back like an undertow.

But much of the air is high enough by now to have an expansive view of the heartland: the lakes, the quilted farms, the crushed stone of old mining territory, the scattered herbicide factories. There is also the curlish smoke from woodstoves. It points east and north—like the weather vane.

No one is quite certain what all is in the wind from California, or for that matter in the currents from Oregon and Washington, from New Mexico or Colorado (where, among many other things, the air has been spiked with plutonium).

While the overall trend of air is eastward, it can move every which way, and in the heartland (which we will look at first because it is the nation's central collection point), the wind comes not only from the west but more importantly up from the Gulf states. Mixing together, these

3

western and southern air masses pick up the formulations of the Midwest and then proceed eastward.

The central part of America, then, while once thought to be pristine, is rather like a huge pot that accepts what the west and south gives, mixes in its own formidable concoctions, and overflows with a nebulous broth.

Though the winds vary according to their height, and according to heat and terrain, they generally sweep across the entire American landscape, whirly and often unpredictable, joined by a tunnel system from Texas, perhaps, or a whiff of Louisiana oil, the soot and pesticides of the great Bread Basket.

From the Deep South comes the aroma of paper manufacturing.

Up the Atlantic coast, gases and aerosols converge with the New York megalopolis and its own daunting smog. The colors change with the altitude: a cerulean blue smudged with charcoal at one height, and below that, a hint of pink or yellow-lime.

At this particular juncture, before the mass moves on to New England or disappears at sea, there stands, of course, the Statue of Liberty, refurbished after her long ordeal with the corrosive atmosphere.

Like her, America, as it approaches the 1990s, is also renewed. The economy and spirit have been mostly vibrant in recent times, the national will refueled.

And after many years of recklessly squandering its resources—poisoning major river systems, and sinking stinking barrels of industrial rot into the soil itself—America is finally witnessing the evolution of a greater consciousness. The nation has begun awakening to the dire hazards of toxic chemicals, watching, with a collective gasp, as citizens from its every region—fellow citizens who are supposed to be endowed with the unalienable rights of Life, Liberty, and the pursuit of Happiness—have found themselves trapped in contaminated homes or sickened by poisoned wells.

They are the victims of an arrogant mind-set within industry that in the end scorns their right to good health and peace of mind.

Once they realize what is happening and look to their local and federal agencies for help, such victims find, with unnerving predictability, that these agencies—designed to protect public health and the environment—seem more concerned in protecting the offending industry.

The scientists in such bureaucracies do not really know what the chemicals might eventually do—how they might wreak long-term harm—but they hide this ignorance behind a wall of haughty technical jargon.

In short, America's chemical technology has grown out of control. Through the rest of this century the United States (as well as other countries) will remain a nation of so many molecular inventions that no one will be able to keep track of them all. They are scattered about the entire land with astonishing and frightening abandon: pigments, plastics, the pesticides that bequeath a green thumb to anyone who can stand the smell of chlorine and phosphorous.

It is a nation with a fetish for the new and the synthetic, but, unfortunately, a substantial proportion of the synthetic materials are at odds with the basic functioning of the human organism.

I had first seen this underside of technology as a journalist who took part in the exposé of Love Canal. There I had watched as an entire community in upstate New York was forced into permanent evacuation because of the treacherous residues dumped there by a small-minded, money-blinded, out-of-control technology. (That and other horrors concerning hazardous wastes were chronicled in a book of mine entitled *Laying Waste: The Poisoning of America by Toxic Chemicals.)*

While this same technology has had a major hand in making America wealthy and great—innovative and fed so very well—it has come to be idolized as some sort of infallible god, when in fact chemical technology is only a tool of man, and a surprisingly limited tool at that.

Its engineers can fasten together sprawling compounds never known before to the earth, yet such technicians lack the means of reversing this process and safely *un*fastening many of these very same compounds.

There are simply no current means to render many of our products harmless. Though not yet conclusive, there is evidence that chlorinated compounds released into the atmosphere, especially chlorofuorocarbons, are depleting the atmospheric shield of ozone that protects us against potentially harmful ultraviolet rays. By one estimate such radiation may cause 40 million cases of skin cancer and nearly a million cancer deaths in the United States during the next nine decades. Damage to the immune system, the eyes, and to crops may also result.

But while science seems unable to reverse or even fully gauge such environmental damage (and has shirked its proper leadership role), the public, in its own equal genius, has turned to another tool of intellect that has proven more reliable through the ages: common sense. The

5

people know that if their lavishly paid doctors have no way of curing a whole plethora of environmental ailments (and if, as in the case of cancer, researchers cannot even prove a "scientific" cause), it is best to play it safe and stay away from *any* quantity of a carcinogenic substance.

Pollster Louis Harris once said that "the deep desire on the part of the American people to battle pollution is one of the most overwhelming and clearest we have ever recorded in our twenty-five years of surveying public opinion." In July of 1984, Harris reported that 87 percent of 1,256 adults he had polled favored stricter enforcement of environmental regulations, including tight control of ill winds such as those that rise from the West or Midwest. He was also quoted as saying that the American public set environmental quality "on a par" with concerns about the economy.

The people know that America really has the ability to correct assaults on its biosphere if it wants to, they know that doomsday is not yet at hand, and they know, unlike their political leaders, that the cost of preventing any such doomsday will not be nearly as prohibitive as industrial lobbyists would have us believe.

Anyway, the cost of saving lives, according to Americans, is never too high. When another pollster, the Gallup organization, ran a similar survey in 1984, it found 61 percent of Americans placing a higher priority on environmental protection than on sustained economic growth.

While my previous focus had been on the poisoning of water and soil, I had also been set to wonder through the years (and through visits to 130 or so of the continent's larger urban areas) about what might likewise be finding its way into the communal air—and thus into those winds that the weatherman tells us move most nights from the south and west to the east.

And I was concerned about much more than the traditional, smog-type pollutants.

Put simply, were the same chemicals that raise alarm if they are found in fish or tap water—benzene, PCBs, DDT, mercury—also in the air we breathe?

We know from episodes of acid rain that pollution can travel over very long distances. Lakes in New York and Canada are feeling the effects of coal-burning in the Ohio Valley. The key culprits are sulfur and nitrogen, and it is certainly a shame for the fish that go belly-up in those acidic lakes. Such fallout also has to have some kind of adverse effect on human lungs.

But I wondered about pollutants that might have a more serious, unstudied effect on humans than mere acid. Were cancer-causing solvents around? Had insecticides in the class of DDT escaped into the general atmosphere? Were there communities breathing *dioxin?*

Such questions seemed more relevant after the chemical disaster in Bhopal, India. In that case a release of methyl isocyanate caused two thousand people to mortally suffocate, and their collective image is etched into our technological conscience: screaming, blinded old women running to nowhere, holding on to fading children whose lungs had been seared.

Since Earth Day in 1970 there has been a great deal of attention paid to environmental issues, and air pollution has always been prime among them.

But even at the pinnacle of its prominence—before it was overshadowed by the issue of groundwater contamination—concern about air pollution was addressed in a very narrow fashion. The focus was on only the six most common and widespread pollutants: carbon monoxide, ozone, sulfur dioxide, particulates, nitrogen oxide, and lead.

While these are extremely important to regulate, there are many other substances that work their way into our supply of oxygen and in the end may be just as damaging—if not more so. By July 8, 1986, the American Chemical Society had registered 7,936,191 chemical compounds, and though not all are toxic, and not nearly all are in common use, a good number of the most toxic ones can be found almost anywhere.

Yet, of these, federal air laws in 1987 controlled less than one in a million—a grand total of seven.

The potential for the others to reach our air passages is a frightening thought. We have done well cleaning some of our vent pipes, we have gotten rid of the unsightly plumes of thick, black, particulate smoke.

But now it is time to cast attention onto the less visible, more exotic threat of airborne *toxic* chemicals, the same kind that were found in the soil at places such as Love Canal.

Unnoticed, they are wafting from our factories, our storage tanks, our incinerators and diesel engines—even from the neighborhood dry cleaner.

And they are spreading through our global village like a toxic cloud.

2

In the 1960s and 1970s scientists equipped with new analytical tools began identifying an array of mysterious synthetic compounds in the flesh of trout and other fish caught in the largest of the Great Lakes, that 31,700 square miles of fresh water known, with due respect, as Lake Superior.

PCBs and DDT were among those found. So was another potent chlorinated insecticide called toxaphene.

In Superior's Whitefish Bay, baby birds began suffering from cataracts and edema. Their necks and heads would become so swollen they could not open their eyes.

Nor could some of them eat. They suffered from a defect that prevented their upper and lower bills from meeting. It was called "cross-beak syndrome."

In other parts of the Great Lakes, especially near Green Bay, the problem was more acute and would begin spreading. In some spots terns soon would be observed with clubfoot-like deformities that prevented the birds from being able to stand. Mink and river otters were disappearing from the south shore of Lake Ontario and around Lake Michigan. Though they had been making a comeback, bald eagles, by the mid-1980s, would have trouble nesting along Lake Superior, and the problem, it was suspected, was that the birds were eating chemically "hot" gulls.

There was also concern about human infants along the lakes. It was feared they might incur learning difficulties because their mothers had dined on contaminated fish. In one part of Michigan test babies seemed to have lower birth weights and smaller head circumferences when compared to those born to women who ate no fish. Some of the babies had what appeared to be abnormally jerky and unbalanced movements, weak reflexes, and general sluggishness.

While Lake Superior's vast shoreline could be expected to receive chemical runoff from farms and scattered industry, the levels and character of its toxicants did not quite fit with what would be expected to enter the lake through the sewers, creeks, and rivers that empty into her.

Located at a relatively remote region between northern Michigan and southern Canada, Superior had not been plagued by the same problems that historically befell its little sister, Lake Erie. Erie's water had been starved of oxygen by phosphates which formed suds at outfall pipes and inspired wild overgrowths of algae—causing bloated fish to wash ashore.

That pollution had taken no modern technical gear to detect: On Cleveland's Cuyahoga River, fires had ignited the oil slicks into a biblical spectacle of flames shooting up from the water itself.

But several hundred miles to the northwest, Lake Superior still had the aura of purity, and well she had better. Nearly the size of Indiana, the lake serves as headwaters for the greatest freshwater system on earth, a system that contains 20 percent of the planet's fresh surface water.

Yet suddenly, in the 1970s and 1980s, its unsullied image was in great danger of forever being shelved. The compounds being found in Superior were mostly invisible and odorless in nature, but they posed much more of a danger than sulfides, raw sewage or oil. Toxaphene possesses the widely recognized capability of causing thyroid carcinomas in rodents; and though not quite as persistent as DDT, it too accumulates in ecosystem, it too is biomagnified in animal flesh, and it is every bit as toxic as DDT, if not more so: One occupational standard allows three times as much exposure to DDT as to toxaphene. Its concentrations in the lake approached levels at which consumption of fish is banned.

The pressing question: Where was the toxaphene coming from?

In Canada it had been closely restricted by government regulators, and on the American side its major use was not near the Great Lakes nor anywhere *else* in the Midwest but instead in two impossibly distant areas, California and the Deep South. The war it waged was against the boll weevil, on cotton fields.

Had it somehow become a constituent of the ambient air?

Had it descended like acid rain or nuclear fallout?

Looking to test an even more isolated ecosystem, scientists began

10

journeying to Siskiwit Lake in a national park called Isle Royale. The island is situated in the northern part of Lake Superior more than thirty miles from the nearest population center of any real proportion, Thunder Bay in Ontario.

To more fully appreciate the virginal qualities of this territory one should know that the only means of access from one part of Isle Royale to another is by foot trails.

There were no outfalls, no farm runoff, no toxic dumps. Certainly there were no cotton fields.

Nor did any of the tainted water of Superior flow into Siskiwit Lake, for Siskiwit's elevation, propped as it is on the island, is nearly sixty feet higher than Lake Superior's.

All things considered, it seemed, there was no way at all for toxaphene to get into Siskiwit. The inhabitants were not boll weevils but moose and ducks and balsam.

Yet to the shock of the investigating scientists, fish netted from Siskiwit had nearly *double* the PCBs that had been found in Lake Superior, and nearly ten times as much DDE—a breakdown product (or "metabolite") of DDT.

One might rationalize certain levels in Lake Superior, with its vast shore and its exposure to at least some modern effluents.

But, now, how did the stuff get to the isolated environs of Isle Royale?

Scientists hauled their gas chromatographs to various parts of the country searching up clues. Again, eyes turned to the distant cotton fields. One big hint was found when rain was tested at an estuary in South Carolina. Toxaphene was found in more than 75 percent of the samples. It was apparently lifting off the fields and into the wind, then raining down.

Perhaps more vital were the tests from the state of Mississippi, another major source. In Greenville, on the lower Mississippi River, there were levels in the air of 7.39 nanograms per cubic meter.

Tracking it northward, air samplers also documented its presence in St. Louis at 1.18 nanograms, and up in Michigan at .27 nanograms.

For those who thought such compounds remain firmly earthbound, or that what little bit does become airborne would simply disappear in the troposphere, the numbers were startling ones indeed. By the time it was 825 miles from Greenville and approaching the neighborhood of Lake Superior, the toxaphene was still at 4 percent of the Greenville level.

At the same time, toxaphene was also being tracked over Bermuda and the North Atlantic; in the fish and water birds from Swedish lakes; in the North and Baltic seas; in the Tyrolean Alps; and in Antarctic cod.

The conclusion seemed as obvious as it was momentous: We were no longer talking about the long-range transport of just sulfur and nitrogen but of the dreaded chlorinated pesticides—thought previously to be a crisis only in lake and river sediments or toxic dumps.

Chlorinateds, the compounds Rachel Carson *(Silent Spring)* worried about in the soil, were taking wing! In quantities great enough to threaten distant wildlife and people!

No one was taking a bullhorn to the streets with such blaring news, but, soon, chemists tapping their computers at the University of Minnesota would start making somber estimates that would have seemed like sheer nonsense just a few years before. Perhaps 85, perhaps 90, percent of the total PCB input to Lakes Michigan and Superior was not from sewers but through the air, they said. And as much of it fell in *dry* weather as when it rained. The number of chemicals in the lake system was estimated to be somewhere in the range of four hundred to eight hundred.

Whatever the number, it was a bad signal not just for cormorants and herons but also, perhaps, for the 26 million people who drink from the lakes. One 1981 estimate said about a million pounds of polycyclic aromatic hydrocarbons, a dangerous sort of toxics, were falling into the lakes each year, along with substantial quantities of benzene hexachloride and DDT.

The DDT, however, looked relatively *new.* It wasn't the old stuff which had been banned in the U.S. for a decade and would have broken into its metabolites by now.

Where from now?

Scientists conjectured the DDT was coming over the poles from parts of Europe and Asia where the pesticide still finds heavy use.

Mexico is another probable source, along with other countries south of the United States border.

In tests in places like California, the results showed that the insecticides did not necessarily need a particle to ride upon but instead vaporized in great quantities right off the plants themselves. One study reported a 59 percent loss of toxaphene from a cotton field within twenty-eight days. Other experiments in closed agrosystem chambers indicated that 24 percent of toxaphene turned into gas within ninety days of the last application, meaning that, in 1974, at least 142.5 million pounds of it vaporized into the American atmosphere.

That meant we have been directly *breathing* chlorinated pesticides and PCBs, not just eating them with contaminated fish. Once the emotional reaction subsided, more questions arose. For example: If toxaphene and PCBs are so airworthy, what else is in the wild blue yonder?

I remember seeing not just cotton in Mississippi but also gypsum and open-waste burning and brown smoke from a furniture factory. There were plastics in Gulfport, titanium in Biloxi; in Pascagoula, the potent nitrobenzene.

More foreboding is the fact that ozone from neighboring Louisiana had been detected at Port Bienville. There was even a town in southern Mississippi called "Ozona," and another where rubbery wastes hauled in from Louisiana were bubbling to the surface—known to the residents as "The Valley of the Blobs."

Were residues from any of *this* accompanying the northbound toxaphene?

Of more immediacy: What was going on in other parts of the Midwest—in the nation's central stirring "pot"—that might also shoot poisonous molecules into the air and then the watershed? And what was happening to those who are most at risk: the people who live close to factories, incinerators, and other sources of materials that are more toxic than simple monoxide or sulfur or even toxaphene?

3

Three hundred and fifty miles south of Siskiwit Lake, in old lumbering and farm territory where the Chippewas once roamed, there is a scattering of tiny hamlets that are barely distinguishable from one another, all of them middle-class or less, all of them past their better days. They are connected by drainage ditches, flat fields of sugar beets, and an unengaging gridwork of two-lane roads.

To hear the locals tell it, they also share an old-fashioned ghost story. Settled 149 years ago, as soon as the government had driven off the Chippewas, this part of Michigan was cursed from the beginning by sandy, gravelly soil, and if there were some healthy plots of berries and barley, along with corn and the sugar beets, the earth was not nearly as generous as it was in other parts of the Midwest. The settlers were forced to search far below the surface if they wanted something other than lumber to sell.

There, in the depths, they found some coal and some oil but mostly what is called Sylvanian brine—an ancient sea of cloudy salt water.

The brine was almost worthless, a threat to fresh wells and quite alien to healthy crops. But it had found at least one aficionado in a Cleveland inventor named Herbert Henry Dow. Within the salt water, he knew, were high levels of bromine, chlorides, and other chemical elements. Bromine was needed by the fledgling photographic industry, the chlorides for bleaching powder.

Dow hit on the idea of extracting them by passing an electrical current through the brine. The charge polarized the minerals and caused them to precipitate. It was modern alchemy, this time with iron mesh, rods of carbon, and tarwood cells.

15

There were of course several false starts. Waiting, like a mad scientist, for the materialization of black-speckled crystals, which would indicate bromine, he instead suffered the indignity of explosions that wracked his first chlorine cells. The roof of one building collapsed and its sides blew out.

Nonetheless, "Crazy Dow," in storybook fashion, persevered, and by the turn of the century, when the mainstay of the area's first industry —timber—had all but disappeared downriver (leaving behind fields of unsightly cork-pine stumps), Herbert Dow had established a company that seemed to know no bounds. Since the official incorporation of the Dow Chemical Company on May 18, 1897, the company has manufactured literally hundreds of agricultural products, dyes, plastics, basic chemicals, and pharmaceuticals, not to mention the many other compounds—many *toxic* compounds—it has handled as intermediates.

Each year, it seems, molecules were latched and cross-latched into ever-longer chains: If the country needed rubber, the company helped make synthetic rubber; if magnesium was needed for planes and incendiaries, Dow provided the magnesium. A special blessing in wartime, it also made mustard gas.

After the chlorine, Dow's brine begat carbon tetrachloride which begat chloroform which begat, according to historical accounts, a rash of rusting utensils and vomiting in the households downwind.

Setting a pattern that would guide the company through years of similar (and increasingly serious, even bizarre) problems, Herbert Dow had in effect muffled complaints with his stirring contributions to the growing economy of Midland, the company's headquarters. Also, the Dow family or related foundations donated land and money for a stadium, an arts center, and residential subdivisions. The courthouse, in large part, had been financed by the old man.

And Midland, despite its modest size, and thanks to Dow, had its own symphony.

"Dow Midland," the vicinity might just as well be called. The very emblem for the city of Midland is the skyline of a chemical complex along with the basic tools of the trade: a tube, a pipette, and a laboratory flask. And at the nearest airport, above the luggage belt, is a sign that says, "Welcome to Dow Country."

The mayor in this town of 38,250, Joseph Mann, is an analytical chemist at Dow, and two of the other four councilmen also draw their paychecks there. So do 7,101 others who live in the area.

Serving as flagship for a corporation that now ranks as the nation's second largest chemical producer (behind DuPont, with annual sales one recent year of $9.5 billion), the Dow facility in Midland is a sprawling one by any measure, consuming nearly one seventh of the city's acreage.

Through the years, when the worshipful townspeople glanced toward Dow's ever-growing, ever-inventing complex, the vent pipes looked more and more like church minarets, and the smoke, in the jargon of a factory town, "smelled like money."

If Dow would eventually gain huge fame for products as diverse as napalm, Saran Wrap, and Styrofoam, a more basic and yet less celebrated turn in its history was the development of a process that made phenol.

An important raw material for plastics and explosives, phenol is based upon a six-sided ring of carbons that would be identical to the famous solvent benzene save for a hydroxyl group (one hydrogen, one oxygen) that clings, like a big crystal on a chandelier, to a handle on the hexagonal ring. The five other handles are occupied by single hydrogens.

Sweet smelling like other derivatives of benzene, phenol is thus known as an "aromatic" compound, and its basic structure, a hexagon like benzene's, is thus known as the "benzoid" ring.

Because phenol contains both hydrogen and carbon, it is also part of the nearly limitless category of carbon-based chemicals known as "hydrocarbons."

Phenol was made at high pressure by reacting chlorbenzol with another product of electrified salt water: caustic soda. Besides its other uses, the end product has long served as an ingredient in textiles, synthetic resins, and as an antiseptic in pharmaceuticals such as skin cream.

As a sort of surprise bonus, two of the by-products of phenol manufacture turned out to be new, never-before-seen, polysyllabic compounds that became the active ingredients for effective (and thus marketable) fungus killers and insecticides. One was known as orthophenylphenol, the other paraphenylphenol.

By 1936 (six years after Herbert Dow had succumbed to liver disease) chlorine and phenol were regularly wed to form products that preserved wood from fungus and mold. They too were given a logical generic nickname: "chlorophenols." By the end of the next decade variations of the basic mixture made for potent weed killers.

It was during this frenetic, can-do, damn-the-torpedoes time, how-

17

ever, that the haunting of central Michigan began. In 1937, as an un-heeded omen, twenty-one workers in Midland had developed unusual skin irritations and rashes after working near the chlorophenol process that made the wood preservatives. The disorder would come to be called "chloracne," since the chlorine products appeared to be involved.

Another eruption occurred in 1964, again near a chlorinated phenol process that made a variation called "trichlorophenol," used for potent herbicides. The "tri" meant each molecule had three chlorine atoms attached to a benzoid ring—replacing hydrogens. The "chloracne" meant a miserable array of symptoms: blackheads, bumps, boils, lesions, pustules.

It started behind the ears and spread down the face, sometimes all the way to the scrotum and back of the legs.

Sometimes the skin turned purplish. Other times it bronzed.

Among the chlorophenol products that could cause such symptoms were the herbicides 2,4-D and 2,4,5-T. Combined together in an oily mixture, they formed the product code-named "Agent Orange." Like many Dow products, it found its spotlight in wartime—in this case during the war in Vietnam, where it was sprayed from planes to destroy jungle canopies under which the guerrillas hid.

Midland did not seem to mind that this powerful agent was being manufactured in the neighborhood, but there would be screeching pro-tests elsewhere, especially among Vietnam veterans who had been ex-posed to the spraying and now claim, along with Vietnamese, that the defoliant—a trichlorophenol herbicide that was made at breakneck speed during the 1960s—caused cancer and birth defects.

This was another ominous sign that Midland should have paid atten-tion to, inasmuch as the homegrown company supplied 32 percent—millions of gallons—of the Agent Orange.

More importantly, because the volumes would be even greater, the chlorophenols were used across America not only as wood preservatives but also, as in Vietnam, to kill unwanted brush and other such vegetation (in this case vegetation that encroached upon electrical lines and high-way rights-of-way). At least one such product, known as silvex, was also widely used as a weed killer on home lawns.

Just as other by-products had formed during the original manufac-ture of phenol, so too was a new, unexpected compound created during the production of trichlorophenol. This by-product looks on paper like four chlorine atoms dancing a square dance with two benzoid rings (connected by a couple of oxygen atoms) in the middle.

Because its name is twenty-eight letters long, prefaced by four iso-meric digits, the compound is kindly abbreviated to "2,3,7,8-TCDD"—an acronym that rings fear into the normally sedate world of toxicology. For if TCDD's acute toxicity to the guinea pig is any criterion, this compound—which the public knows as "dioxin"—is the most toxic synthetic chemical known to man, far, far more toxic than toxaphene, or DDT, on a par, nearly, with nature's supreme poison, the botulinum toxin.

The scientific background, however cumbersome, is a vital subplot in our mystery. For instance, it was reasonable to presume, said a 1980 report by the Industrial Environmental Research Laboratory of the U.S. Environmental Protection Agency (EPA) "that the slightest trace of 2,3,7,8-TCDD in the environment may have adverse effects on the health of both humans and animal populations."

In plainer language, TCDD (or "tetrachlorodibenzo-para-dioxin") is a big problem as soon as it can be detected. I had seen this type of dioxin cause official alarm at levels that would tax even a microbiolo-gist's imagination: At the Love Canal, an uproar was caused when dioxin was found in mere parts per trillion of liquids seeping from that old dump—that is, several molecules of dioxin for every 1,000 *billion* mol-ecules of groundwater.

And that hysteria was over dioxin that had not even been detected at the surface yet. It was simply buried there in the ground. The federal government had decided that if dioxin was found at more than one thousand parts per trillion (or one part per billion) in surface soil where people lived, that was cause for emergency considerations. Indeed, the entire town of Times Beach, Missouri, had been evacuated in 1983 as a result of dioxin levels in this range.

A part per billion (to bring home a point that needs constant bring-ing home) is comparable to but a single inch in sixteen thousand miles.

In the trichlorophenol residues at Midland the level of dioxin was far higher than in the finished herbicidal products—up to 10 billion parts per trillion, according to information Dow itself provides.

How much dioxin of the awesome 2,3,7,8-TCDD isomer Dow had created cannot be precisely discerned. But no one can doubt that, con-sidering its toxicity and the great volumes of chlorophenols the company produced, the quantities had to be truly daunting ones.

19

So were the potential effects. Though there remains a paucity of data on how dioxin affects humans (few people are willing to act as test subjects), some of the potential ramifications can be drawn from occupational exposures involving TCDD, along with the more tenuous conclusions based upon experiments with not only the pitiable guinea pigs, but also hamsters, rats, frogs, and dogs.

According to one chemist, Philip M. Cook of the EPA's environmental research laboratory in Duluth, Minnesota, dioxin is seventy thousand times as toxic to mice as cyanide. In guinea pigs, the amount that would cause immediate death is probably on the order of one hundredth the size of a grain of salt and perhaps even smaller.

But it is not the acute effects that worry most scientists. Rather, they fret over the long-term—or "chronic"—exposures.

And they wonder if dioxin can cause trouble at levels they don't even have the technology yet to detect. There are those who say tumors or other ill effects in animals have shown signs (as during one University of Wisconsin study) of starting up with doses of dioxin at such levels as five parts per trillion (that is, five inches in a stretch 16 million miles long—about sixty-seven trips to the moon—or, as one company more palatably calculated it, a quantity comparable to about five drops of chocolate in 1.28 million barrels of milk).

In some laboratory specimens, dioxin seems to target the liver and thymus. Test species also have been shown to develop dry, scaly skin and lose their hair. Monkeys have experienced abortions, mice developed kidney abnormalities, and golden hamsters were born without their eyelids. A common cause of death in the guinea pig is a simple wasting-away syndrome.

In the view of the trichlorophenol producers, many of the studies warning of such powerful biological phenomena were badly flawed. The tests were based on too small a study group, the statistical method was biased, or the doses of dioxin were too high—you could hurt an animal with anything, if you gave it too much.

Or, the rat strains were unusually susceptible to such poisons. That, or perhaps their tissues had been wrongly diagnosed.

If the tobacco industry had been able for so many years to dispute evidence that cigarette smoking causes health disorders, it was going to be even easier for the chemical industry to keep a controversy swirling around something as complex, subtle, and often intangible as the white crystalline substance called TCDD.

When I asked Dow its standpoint, the company graciously sent me

20

a box of scientific documents from around the world. Included in the impressive response was a technical report by the American Medical Association. "The studies to date on the human health effects of Vietnam exposures to Agent Orange do not reveal a clear relationship between serious illness and exposure," it concluded, saying that despite all the effects in animals, chloracne was about the only definite effect in man. More studies, it said, were necessary.

Companies such as Dow were very correct in pointing out that in some cases humans seemed to be thousands of times less sensitive to dioxin than certain laboratory animals, and perhaps equally right when they argued that TCDD had caused unnecessary panic in many parts of America.

There are many who believe that the world's press has overemphasized the threat of dioxin, for, as we shall see, there are some equally arcane molecules out there that are very close, in potency, to dioxin. Other toxic materials are much more widespread.

Dioxin, however, has a short, sexy ring to it, and there is even an *x* in the middle syllable that brings forth images of the skull and crossbones on a classic poison bottle.

The various dioxin accidents at factories have yielded no "whopping outbreaks" of any particular fatal disease, noted Linda Birnbaum, who tests for the effects of dioxin at the National Institute of Environmental Health Sciences (NIEHS). But as Dr. Birnbaum also noted, there is as yet no proven threshold—or "switch"—below which this type of chlorinated dioxin (there are seventy-four others) has proven to be harmless. Scientists are still looking for the final "no-effect" level. Even after the federal government alone had spent more than $1 billion looking into it, there were as many questions as answers.

Suffice it to say, however, that if animal tests are any indication, TCDD can affect about any organ in the body. As a carcinogen, it possesses almost unprecedented potency.

When scientists search for a benchmark of dioxin's human effects, they often refer both to the Vietnamese exposures and also to an accident on July 10, 1976, in Seveso, Italy, in which a valve on a trichlorophenol reactor released dioxin of the TCDD type (which we will represent in the singular, as opposed to other, less toxic "dioxins") and led to the evacuation of three hundred acres. Reports persist that the exposure caused not only chloracne but also spontaneous abortions and certain birth defects, among them spina bifida, neural tube defects, and polydactyly, the scientific term for a person born with too many fingers or toes.

While there is still no final conclusion on whether dioxin is actually teratogenic (a cause of birth defects) in man, even the skeptics of dioxin's effects concede that it has caused abnormalities in the offspring of exposed mice at dosage levels ranging from one to three nanograms per kilogram of body weight. A nanogram, it is important to know, is but a billionth of a gram, and it takes twenty-eight grams to make an ounce.

In humans, dioxin has also been blamed for "pins and needles" sensations in the feet, enlargement of the liver (hepatomegaly), and abnormalities in the immune system, which might open a person to any number of infections.

There were also aches in muscles and joints, digestive disorders, nerve changes, and even psychiatric effects—including the hearing of phantom voices.

4

In 1979, in one of the tiny rural hamlets near Midland—the German community known, with chilling happenstance, as Hemlock—rumors began circulating; rumors of inexplicable and gruesome deformities afflicting the animals.

It was the phantom that farmers claimed roamed from borough to borough. From house to barn.

A biological gremlin! Rabbits grew large tumors, supposedly, and a chicken three legs. In one barn, mice were seen running in dizzy, epileptic circles.

Cows and horses had mysterious sores and lost their fur in clumps. Or so it was said. They seemed to be swelling in the extremities, or simply wasting away.

In the wilderness were deer with green venison or white liver spots.

If one neighbor appeared to have won the freakstakes by claiming that *both* chickens and cattle had blistered livers, that was actually little in comparison to a couple of geese I saw in a pen behind another home. They had been born with their wings on backward.

While health officials and others scoffed at the legend, finding little evidence of any kind of plague, a growing number of residents, especially a weathered farmwoman named Carol Jean Kruger, insisted that the curse was also claiming humans. There were kidney problems and bone disorders and cancer.

She theorized—perhaps correctly—that the tap water was to blame. Her own toenails had turned brown and her teeth had broken off at the gum line.

In an age when the environment and the chemicals in it had become so suspect—especially in Michigan, where thousands of eggs and live-stock had to be destroyed after an accident in which the flame-retardant PBBs (polybrominated biphenyls) had been mixed with farm feed—not a few eyes, in Hemlock, were turning in search of an answer to that historical benefactor in Midland, the Dow Chemical Company.

But this time the worried eyes were no longer so worshipful. Now, when they looked at the smokestacks, their glint was accusatory.

Unthinkable in previous decades, there were even harsh letters criticizing Dow in the local newspapers.

Though, originally, Dow had declined to talk to me about the situation, and though, more recently, the corporation had suddenly canceled a long-standing appointment I had with one of its board members and vice-presidents (after I had sumbitted what may have been wearisome questions), the firm, by 1987, was cordially attempting to provide answers to potentially embarrassing questions.

The view of Dow's president, Paul F. Oreffice, who also declined to meet with me, could be gleaned from interviews or guest columns in the hometown and Detroit press: "The air of Midland is our air; the Titta-bawassee is our river. The vast majority of us don't come from some other community and put in our workday and then go back to that other community, leaving behind our concerns for its well-being.

"Midland is home. That counts for a great deal. It accounts for the fact that we want to take pride (and do take pride) in the environmental quality of Midland and its health record and its safety achievements."

"People tend to exaggerate and try to run you out of business for no good reason," he said another time. "I've said it before and I'll say it again: In humans, dioxin has not been proven to do much more than cause a skin rash."

Because it is not a controlled setting, there are enormous problems investigating perceived health effects in an actual community. People move in and out. There are no regulated doses. They are exposed to many different substances, a good number of which cannot be recorded precisely.

The result is that virtually never is there any irrefutable proof that a disease—if it is at abnormal levels, which in itself is extremely hard to prove—has been caused by any one factor, such as a particular chemical.

In Hemlock, the local veterinarian told me he had not seen any anomalies consistent enough to suggest a plague. Scientists, especially state scientists, reminded me that a certain percentage of genetic quirks were to be expected anywhere: About one percent of laboratory mice are traditionally born with serious birth defects, for instance.

Other problems could have been from inbreeding, poor feeding, and bad hygiene in the barn, which would invite any number of parasites and bacteria.

Carol Jean's skeptics also pointed out that when the community's well water was tested, chemicals appeared, for the most part, in only "insignificant" amounts—comparable to what would be found in the modern environment nearly anywhere.

I returned to the Midland area in 1986 wondering three things: whether the weird symptoms were still being reported, whether they went beyond tiny Hemlock, and whether, from all the local activity with chlorophenols, traces of dioxin—that's all it might take, traces!—had gotten into the air.

If so, it might explain some of the symptoms there.

For years America had dreaded the chemical contamination of its water resources. Or of food. But few had looked for it in the air, a medium where it can travel the furthest.

Again, I am not speaking about the classical pollutants such as sulfur dioxide and carbon monoxide.

Instead, I am asking if there were health mysteries in Midland that might be related to chlorinated and benzoid compounds causing not standard acid rain but *toxic* fallout.

I paid a visit to a short, comely woman named Diane Hebert, who lived very close to the Dow complex. Reaching middle age, with penetrating bluish-green eyes, she was a dynamo of energy who had worked at a number of jobs in her life, including one as an airline stewardess. Her own husband worked for a local Dow subsidiary as a corporate pilot.

Diane had moved to her house in 1977, and the odors got to her right away. "You'd sit there and watch TV and you'd smell it, especially during weekends, especially at night," she said. "Sometimes it was like an ammonia solution, sometimes like chlorine bleach, sometimes like sulfur. I didn't know what to do about it. My son gets bloody noses, and it seems like it's after these smells."

In Dow's chemical process there were 1,950 potential "point"

25

sources of emission, mainly relief valves and vents, according to a state official who spoke only after I agreed to go through the Freedom of Information Act process. A "point" source emits pollutants from a single stationary location, while its counterpart, "area" sources, are collective groups of relatively small, mobile, or intermittent sources, such as the local dry cleaner or an automobile or a kerosene burner.

The figure of 1,950 didn't count regular smokestacks (another example of "point" sources) or wastewater systems, which contain significant levels of various volatile organic compounds—chemicals that readily evaporate, like alcohol.

Nor did it take into account dust blowing from contaminated soil.

As for the incinerator, in 1983, I learned, it burned 64.8 million pounds of various waste in a rotary kiln, which evenly and intensely heats it by turning it around like a cement mixer. In addition to whatever may already exist in such wastes, TCDD and other dioxins can be formed anew by the heating of certain industrial materials in a heated unit.

Diane lived, in a sense, under the incinerator's plume. And if, in response to the factory's odors, she was at first baffled as to a course of action, she subsequently steered with great self-assuredness into the role of an ardent citizen-activist. Besides protests, she filled her days collecting reports of neighborhood illnesses.

There was a good deal of the detective in her. And she was *obsessed.* She began collecting all kinds of information about Dow, ranging from little rumors whispered to her by Dow employees to voluminous scientific documents on the nature of chlorinated chemicals.

She believed that her own problems were somehow connected to those in Hemlock, and that if the water was indeed a concern, there was also a lot to ask about the air.

"I had very hard lumps on my back, and a friend of mine from the Hemlock area suddenly had acne and never had it before," she said, sighing and settling into a chair, not knowing quite where to begin, smoking cigarettes, as if the dangers of smoking no longer mattered.

"Another had what he thought were hives around his belt line. I heard when the wind was coming from the east, people to the northwest of the plant got the 'flu.' And I'm in the prevailing southwest wind, so I get regular emissions from the Dow incinerator. Or any other releases."

Along the dioxin trail are many other potential air contaminants to consider in Midland. When, during a trip to Washington, I came across a March 7, 1985, letter from Dow to Gerald P. Dodson, counsel to the House Subcommittee on Health and the Environment, it was attached to an inventory in which Dow reported the annual release of 76,000 pounds of benzene, 366,000 pounds of toluene, nearly a ton of vinyl chloride, and 4.6 million pounds of methylene chloride.

It is important to keep our chemicals straight. Both benzene and toluene, because they often find use as degreasing or dissolving agents, are known as solvents. An important industrial solvent now for more than eighty years, benzene is present in coal, petroleum, and a truly prodigious number of synthetic products, used as it is in the manufacture of other solvents, pesticides, and pharmaceuticals, including the phenol agents.

Clear, colorless, and highly flammable, its structure is based again upon the hexagonal ring of carbons, with hydrogens hanging onto the outside for dear life. In 1985, 9.73 billion pounds of benzene were produced in the United States, or more than forty pounds for every citizen.

Many solvents are immediately recognizable because of ending with that "-ene." Toluene is another benzoid compound—that is, six carbons compose its central ring—and though it is less toxic than benzene, which was one of the very first recognized human carcinogens (a cause of leukemia), they are very closely related. Both are volatile organic compounds, or VOCs—organic in the sense that, as hydrocarbons, they are based on carbon, which is the foundation for life itself. They both should also be classified as "aromatic," implying the presence of a benzene ring.

Take the benzoid ring, add a carbon and extra hydrogen to a side chain, and, like a rabbit from a hat, you have sweet-smelling, glue-sniffers' toluene.

Though not recognized as a carcinogen, toluene, which like benzene is readily found in gasoline (as well as paint, varnish thinners, and adhesive products), can cause upsets to the nervous system. Like dioxin, both toluene and benzene can also irritate the skin.

Their important effects are the chronic ones. Compared to some compounds, benzene is not that acutely toxic. In rats the lethal dose is measured in grams. Dioxin, in stunning contrast, is spoken of not in grams or even hundredths of a gram but instead in *millionths* of a single gram (or "micrograms").

As for the others, vinyl chloride, which is used to make plastic, is a proven carcinogen that's at least as poisonous as benzene and probably more so; methylene chloride, another solvent found in paint removers, is substantially less toxic than either of them.

It is difficult if not impossible to know what these materials can do

27

when inhaled together, but we do know that toluene can alter the metabolism (or breaking down) of other solvents, delaying their excretion from the body.

Dow told Michigan's Air Quality Division that five dozen different categories of compounds were released into the air, and though most were not highly hazardous (and were in the process of being reduced), the emissions, by Dow's reckoning, came to more than 48 million pounds a year.

Diane Hebert, housewife-activist, was well aware of such volumes. She nodded toward the phone. "Someone will call and ask, 'Do you think my baby's birth defect could be related to this or that?' Or *cancers.*

"At this point I'm really convinced there's a correlation but I'm not the one to give that kind of news. There are cleft palates, urogenital birth defects: It keeps me up at night. I get in such funks that I don't even want to leave the house."

She said one insider, a former security guard, told her company reports on accidental releases, along with other revealing records, had been altered or destroyed at the plant. (Dow, in response, described that as "a very improbable scenario.")

Diane, however cynical about the company, did not entirely blame Dow for all the calls about cancer. Doing so would have neglected not only the other causes of the disease but also other potential forms of pollution there. While Dow is by far the dominant industrial force in Midland, the general area could also have encountered some pollution from refining and chemical processes in Gratiot County, or from something as basic as regular neighborhood insecticide spraying.

If that wasn't enough to muddle the matter, there was also the fact that, due to the famous accident, nearly all residents in the state had at least some trace of the PBBs in their bodies. Michigan, it was becoming clear, was a witch's brew of contamination.

Still, Dow was by far the likeliest focus for concern. And as a crusader for a group known as the Environmental Congress of Mid-Michigan, Diane had certainly broken the wall of silence that officials up until her time had carefully and tenaciously maintained around the corporation.

Suddenly the executive director of the Midland Chamber of Com-

28

merce, William Welch, felt obligated to attack the environmental movement as "the age of irrationality" spurred by "a small handful of
zealots." (His idea of "quality of life" was the local tennis center.)

An island in a technical maelstrom, with Dow's power surrounding
her, with the city itself openly hostile to her, with the incinerator just
two miles upwind, Mrs. Hebert worked quietly and nearly alone for
years, up until mid-March of 1983, when a controversy suddenly erupted
that seemed to show that Dow had great influence not just with local
and state governments, but with the federal EPA as well.

It was charged that the agency had allowed Dow scientists to influence a government report on regional water contamination.

Called to arms, Diane went beyond simple letter-writing and began
to voice her concerns—and desperation—on television. Just before I
visited her she had also joined the environmental group Greenpeace,
which provided her a modest salary and some badly needed expenses so
the photocopying and postage and phone calls, in waging war against
polluters, didn't all come out of her own pocket.

When the activists from Greenpeace came to Midland to plug Dow's
pipes and, through such a ruckus, shed publicity on the water pollution,
Dow responded with its own brand of hardball. Suddenly Diane was
noticing cars trailing her, the drivers talking into their radios, she said;
and after the pipes were plugged, Dow had the Greenpeacers arrested
for trespass and thrown into jail.

Somehow the company obtained confidential information from the
county jailhouse to the effect that one of the activists, a twenty-nine-
year-old New York woman, was infected with venereal disease.

When a Dow public relations man, Phillip Schneider, called Diane
to let her know he had such embarrassing information ("They were
clearly trying to intimidate me"), the scandal blew into a major local
news story. Even though he had acted on orders from the general manager of Dow's Michigan division, Robert Bumb, Schneider soon after
resigned and the maligned Greenpeacer sued the company.

It was discovered she didn't have venereal disease after all, and Dow
was forced to apologize at a most inopportune time: in the midst of
kicking off a $50 million, five-year advertising campaign to improve its
image (with the now too-familiar jingle "Dow Lets You Do Great
Things").

All of this is in the way of setting a backdrop for what Diane was
now telling me. For if she and her neighbors were just partly correct,

central Michigan was rather like a subtler but perhaps larger version of Love Canal, incurring—or suffering the aftermath of—a major environmental calamity.

Jacquelyn Froehlich, who lived next door to Diane, was equally convinced the air was affecting the neighborhood, and when I asked about the vegetation, remembering, after all, that Dow's products included an impressive stock of herbicides, she said, "Well, we put in tulips and the leaves are all red-streaked. And the tulips themselves are very peculiar-shaped, like dwarfed.

"I can't tell you how many years we tried to grow vegetables. We have an apple tree and it gets beautiful blossoms, but the apples themselves are dreadful—dimples in them and brown streaks. Every single one of our mountain ash trees are dying. And they used to get such beautiful berries! The leaves fell off our sycamore during summer and it looked like autumn. We can't grow grass very well, but that could be from anything. That could be the weather. I'm not an authority on gardening.

"But then there are the dogs. We had a beagle with thyroid problems and it lost all its fur. The dogs around here lost their fur in big clumpy patches and they *itch.*"

While her beagle, Barney, got its fur back after medical treatment, she mentioned that a dog across the street, a Maltese, "looked like a big pink rat." There were four dogs in the immediate vicinity that had thyroid disorders, she said, and Barney eventually developed a mass in its abdomen and a tumor in its testicles. "We put him to sleep in 1984. He had gobs of cysts. And he would stumble and fall. Many of the dogs have arthritis."

Nor was Mrs. Froehlich, a matronly type in her early fifties, worriless when it came to her own daughter, Diane. "She'd go out for walks and have to come back and take a shower, she would itch so much. Every part of her that was exposed—her arms, her neck, her legs . . . She'd be frantic, but if she showered right away the rash wouldn't be too bad."

There was something going bump in the night again, just like in Hemlock years before. Now, ten miles north of there and much closer to the plant itself, the same concerns were consuming the cityfolk, this time pointing not to water contamination but to the potentially far more serious route of the air.

If Hemlock was enough of a mystery to draw the attention of a Senate subcommittee (and warrant an article of mine in *The New York Times Magazine),* then the possibility that the city of Midland itself was sustaining strange afflictions is of much greater consequence, its population far larger than a minor farming community.

Yet, in all the reams of dioxin studies, in all that had been written about Agent Orange, the city of Midland—where so much of the herbicide was manufactured—had barely warranted a footnote.

I exchanged glances with William Gallagher, a television reporter from Detroit who had taken the day off to accompany me. There was mounting surprise in his eyes. Neither one of us had expected quite such detailed testimony from the Midland activists.

I knew many of the problems with vegetation could have happened anywhere, the result of inadequate sunlight, bitter soil, insects, or other natural things like root rot. At the same time, the Midland stories were reminiscent of accounts from the Pinal Mountains in Arizona, where dioxin-carrying 2,4,5-T and silvex had been sprayed on forestland. There, residents had complained not only about dogs losing fur, but also discolored vegetation, deformed or crippled farm animals, hawks that tottered and lost the ability to fly, and dead sycamore branches. Their observations were suddenly being repeated 1,550 miles away, in Michigan.

We were joined by yet two other area residents, Alta Bidwell, who lived three blocks from Diane, and Sandy Mannion, a nurse who lived in the questionable rural area outside of town.

Alta was another who talked about dogs: "The first one we lost was a Saint Bernard. I think it must have been cancer because it got very, very thin. Then our next dog was a Bouvier, a Belgian cattle dog, and I'm not sure how old—it wasn't all that old and it got a heart problem and laid down and died. Now we have a Great Dane and it feels to me like it's got swollen glands in its neck."

Alta's own neck was very swollen—a thyroid problem, it turned out. She wasn't sure why she had it; it wasn't an iodine deficiency, she said.

"There seem to be an awful lot of thyroid problems in Midland. My youngest son gets nosebleeds, like Diane's son, and they happen just zippo, too. We got headaches, and I thought it was the air."

31

Her hair had thinned substantially, and she complained about "phenolic odors." She also worried about the wildlife: there didn't seem to be many birds this year, and where the neighborhood used to be teeming with squirrels, such that she would see at least one a day, the only squirrel she had noticed lately was scrawny, with a short, stubby tail.

And before that there had been something else that had bothered her because it seemed out of the ordinary: a squirrel floating dead in her swimming pool.

Whether or not any of these observations and events had a thing to do with pollution is at best a tenuous matter of intuition. While it is not implausible that a toxic substance could disorient a squirrel (dioxin *does* affect the nervous system), it is also possible that the animal simply fell from a limb or was scared into the water by one of the dogs whose friskiness was still quite intact.

More substantial by far were the filing drawers of documents Diane Hebert had collected in her basement. There were petitions, journal articles, and reports by every conceivable governmental agency. Diane herself had no background in chemistry. Basically she had been the housewife, never making it to graduation at Flint Junior College.

Instead she had gone to a small commuter airline way north in the Upper Peninsula, bookkeeping before her stint as a stewardess, then afterward selling tickets and writing the airline's advertising copy, until she settled herself at home. There were other jobs along the way, but none of them taught her the basic benzoid ring, let alone something called 2,3,7,8-tetrachlorodibenzo-para-dioxin.

But in Diane there were two inner forces at work. One was a determination not to accept the blanket and "knee-jerk" assurances that seemed to come like clockwork from local, state, and federal health officials—as if they themselves were on automatic pilot. The other was the force of simple necessity, an energy that allows a laywoman to learn the twisted intricacies of a technical field when that person feels her family is threatened by that very same technology.

She handed me several sheets of crude, handwritten notes on illnesses she logged in both Midland and its outlying hamlets. There were cases of Down's syndrome, immunological distress, liver disease, brain tumors, respiratory illness, miscarriages, multiple sclerosis, and uterine cancer. All told, she and others were expressing concerns for an area that now involved not only Midland's 27.9 square miles, but more than 100 square miles of central Michigan.

32

She also gave me a stack of recent obituaries from the Midland newspaper, dating back to 1984. On page after page were the faces of people who had succumbed to cancer—either that or, as the obituary writer more softly put it, some other "lengthy illness."

Too many seemed fifty or younger. There was Michael A. Grzenda, who died of a brain tumor at the age of six; there was Brad Lee Hamilton, a senior at H.H. Dow High School; there was three-month-old Samantha Jo Meyers, whose illness, needless to say, could not have been a very "lengthy" one.

The family-album snapshots of these people stood out in haunting fashion, most of them smiling, all of them portraying the wholesome faces of America's middle class. At least a dozen had worked for Dow, and as for the young ones, they took on new meaning when, later on, I saw figures from a 1984 state report showing Saginaw County—home to Hemlock—with an infant mortality rate about 67 percent above normal.

As striking and even lurid as these obituaries were, they naturally fell far short as any form of proof. A certain number of people anywhere can be expected to die painfully young, and when a company employs thousands of the local residents, they are naturally going to occupy a disproportionate amount of obituary space. Dow's own study of workers involved in 2,4,5-T production showed at one point that there were only 11 deaths where 20.3 would have been expected normally.

When I obtained three huge, tan-colored volumes entitled *U.S. Cancer Mortality Rates and Trends, 1950–1979,* I could spot no overwhelming definitive patterns—that is to say, the area in and around Midland was hardly a sore thumb in every cancer category. In Saginaw, the rate for white males between 1970 and 1979 was a bit over the national incidence, and so was Midland County, while the area just north—in Bay County, which is predominantly downwind, in the crotch between Michigan's mitten-thumb and palm—had a rate of 219.2 cancer deaths compared to the national rate of 204.1.

At the very same time, the rate among white women in Bay was *below* the state incidence. This was a potentially important indicator because females would be less likely to find exposure at a factory itself. They don't shovel the toxic residues in a stenching vessel room, they had

33

little to do with the actual mixing, cooking, and packaging of toxicants like Agent Orange. In short, the statistics are not as influenced by occupational-type exposures.

If it was something in the locality's general air that was affecting the people, then it would logically show up among women as *well* as the factory-employed men.

There did seem to be a bit of a blip when it came to lymphosarcoma, and white males in Midland incurred cancer of the nose, nasal cavities, middle ear, and associated sinuses at a rate that was statistically significant (but no higher, to give the reader an idea of the quirks of statistics, than remote areas such as Stillwater County, Montana). In Midland, cancer of the esophagus among white males was lower than the state average and only slightly higher than the rate in Bay County, while the exact reverse was true—Midland was higher and Bay lower—for cancer of the liver and gallbladder.

As far as breast, brain, and ovarian cancers in Midland—often cited by citizens as proof of an epidemic—these were also lower than would be expected in a typical county. The figures for Hodgkin's disease, an ailment often cited in polluted areas across the country, meanwhile, were quite average. There was even a report, admittedly an *incomplete* report, that showed Midland with the state's lowest rate of cancer death.

But at this point it is time for a depressing switch. In Bay and Midland, leukemia—another bellwether in polluted neighborhoods— was higher than the state average for white men. And Midland's rate was above that of Bergen County, New Jersey, where the first famous cluster of leukemia had occurred in Rutherford, near Giants Stadium.

In white women the rate of esophagus cancer seemed, at nearly twice the state rate, to be just another point in the overall statistics where the curve arched upward before it once again fell and averaged out. But there was a striking comparison between the rate in the 1970s and that of the previous decade: the incidence had leaped 324 percent.

Looking back at the nasal-type cancers in men, there had been a 150 percent increase during the same time period. According to yet another set of figures, the number of new cancer patients at Midland Hospital Center was 60 percent higher in 1982 than six years before. Hence, the statistics could be molded into any *number* of mosaics.

In investigating exposures to dioxin, clinicians have long taken a special interest in a rare cancer that afflicts muscles, cartilage, nerves, blood vessels, fat, and tendons: soft-tissue sarcomas.

Much of the concern calls back to 1949, when an accident at a

Monsanto reactor in West Virginia possibly exposed more than 200 workers to dioxin and led to reports—which again are subject to dispute —of soft-tissue cancers among those workers. In Sweden, a survey by an epidemiologist at the University Hospital of Umeå seemed to show that workers handling phenoxy materials had up to a sixfold incidence in soft-tissue cancer, and at Dow itself, in what was known as the 199 Building, where there was the exposure in 1964, the soft-tissue cancer had also reared its head.

Clearly, if dioxin was causing cancers in Midland, it seemed likely that soft-tissue sarcoma would be among them.

When it showed at abnormal levels, it was rather like spotting a ripple in the water that could be the head of the Loch Ness monster.

Once again, women would be a good environmental weather vane, since they did not usually work with dioxin. Between 1970 and 1978, as it happens, white Midland females contracted the unusual cancer at four times the national average. When the National Cancer Institute's 1970– 1979 figures are lined up next to those gathered in the 1950s, the jump in esophagus cancer becomes mild in comparison: Though the total number of soft-tissue malignancies seemed small (at eight) that was a staggering 608 percent jump since 1950–1959. In an internal newsletter dated May 12, 1983, Dow announced that while this type of cancer was more prevalent than it should have been in women (and not only in the 1970s but the 1960s as well), two of a total of thirteen women had been disgnosed before moving to the vicinity.

There was "no link" between the soft- and connective-tissue cancer and dioxin exposure in the county, Dow clearly maintained; a review of death certificates and other data had "failed to reveal any commonality linked to a single cause."

However, there was at least one thing these women certainly did have in common, and that was breathing Midland's air. Was it at all conceivable that the most pernicious of toxic pollutants, the same ghost- like isomer of dioxin that caused a national event because it had been *buried* in a single neighborhood known as the Love Canal, somehow had found its way to a significant degree into the general *atmosphere* of an entire locality?

It seemed nearly too sinister a thought. It was one thing for dioxin to wind its way into a river bottom, another matter entirely if it had spread through the air. In a river, dioxin was certainly cause for deep concern, but it tended to settle out quickly and presented its most significant danger to those, as near Lake Superior, who ate the fish.

For several years it had been known that the river coursing by Midland, the Tittabawassee, was seriously contaminated by dioxin, comparable to the Agent Orange–dosed Saigon River in South Vietnam. And officials at both the EPA and the Michigan Department of Natural Resources naturally assumed that a good share of it was from the wastewater flowing from Dow.

In fact, the only time Midland had been the subject of major national attention with regard to dioxin had come in 1983 during the EPA water-study controversy, when it was charged that the federal government was downplaying Dow's culpability in polluting the Tittawabassee River after allowing Dow to review the report line by line. In the midst of that furor I had recalled a simple line-chart sent to me when I was investigating Hemlock four years before. The graph depicted birth defects in Midland during the 1970s, and the line during that period had risen to a craggy peak that put one in mind of Mount Everest.

It seemed possible that some kind of undetected and perhaps untraceable release—some compound of virtually invisible virulence—had caused a number of extraordinary effects in various spots of these unsuspecting communities.

I stopped at the home of Sandy Mannion, the nurse. Behind her small abode were flat fields of corn and soybeans.

Sandy covered her antiqued kitchen table with a large, homemade map of the rural vicinity. There were colored squares and dots to represent various health disorders in the residents. Dark blue dots were the most ominous, signaling cancer; light blue meant a nonmalignant tumor. Her own house was a pink square, which meant a birth defect or retarded child.

Standing next to me in the kitchen was Sandy's son Clint, ten, a friendly youngster who gurgled all his words, for he had been born with Down's syndrome. "There's another case down the road about three miles, and there was another girl older than Clint who used to live about five miles from here," Sandy explained.

"When he was born I remember a nurse saying to me, 'Boy, we've really had a run on kids with Down's syndrome lately.' That was one of the first things the nurse said to me. There are a lot of retarded kids around here of all ages."

Clint's birth, on February 20, 1976—close enough to that suspect time period of birth defects—brought with it a torrent of understandable emotions in the young mother. "It's hard to pick out one thing," she recalled, nervously giggling. "It was shock, I guess, and some anger, disbelief. I remember that night in the hospital, thinking that when I'd wake up in the morning it'd be gone—like a bad dream.

"But when you wake up it's still there. I cried continuously the whole next day. For a short time I was angry that it happened to me, but that didn't last very long. I figured that the best thing to do was to find out everything I could about it and get going on trying to help Clint.

"He was also born with an enlarged liver and a defective optic nerve and several heart defects. He's had kidney problems—several kidney infections all in a row, which I found out other people's kids around here also had. At first they didn't think he had a kidney on the right side, but then they said, no, they thought it was there but that it was deformed. They never really gave us an answer."

Visits to the doctor caused her to soak in sweat. "I thought that Clint was going to die," she said, explaining that it was trisomy-21 Down's syndrome, a kind that is caused by an unusual division of chromosomes in the ovum or sperm.

She had had a miscarriage a few months before conceiving Clint, she said, and her older boy, Brett, thirteen, suffered skin hemorrhaging and a bone infection. "Pages and pages of illnesses," she sighed.

In early 1987 I learned that Brett had been diagnosed as having a hidden case of spina bifida—a birth defect causing a cleft in the spine— and that the Mannions were waiting to hear whether Clint needed heart surgery.

Sandy had joined forces with Diane to complain about the "yucky odors" that wafted through her windows when the sky was low and the wind was coming from the north. She became increasingly outspoken about the pollution, and suddenly began getting hang-up calls.

5

In most tests of its effects on the chromosomes and fetus, the TCDD isomer of dioxin is fed to test-animals over a period of weeks, months, or even years.

Because I was interested not just in steady exposure but also in what could happen upon a one-shot exposure, which might occur in a place such as Midland during a large but unreported release, I contacted Robert Pratt, head of the experimental teratogenesis section at the National Institute of Environmental Health Sciences.

He told me single doses can indeed cause birth defects, "Most of the time the doses we're working with are designed to produce 100 percent malformations in a certain strain of mice. Most of our studies are on the order of magnitude of between ten to fifty micrograms [again, a microgram being a millionth of a gram] per kilogram.

"That's extremely low, because most of the other chemicals and drugs that produce cleft palate in experimental animals do so at doses that are on the order of milligrams [or thousandths] per kilogram—a thousand times as much. The threshold dose at which we start to see effects in mice with dioxin is around one to two micrograms per kilogram. For our studies we only give it one time, a single shot during gestation."

Interestingly, it is cleft palate and urogenital defects that seem to be produced most readily by dioxin in test animals, especially cleft palate. During certain critical days of pregnancy—what are known in the trade as "windows of development"—dioxin appears to prevent the proper fusing together of two palate shelves by interrupting a cycle in the epithelial cells that initially separate them. That is, the epithelial cells don't

39

die as normally they should, making way for the two palate shelves to join.

That a single dose at such low levels could cause animal malformations meant that there very well might have been birth problems in humans near Midland if dioxin of the TCDD kind indeed had been released into the atmosphere there as either a single, heavy puff during a factory malfunction (as in Italy), or during incineration of dioxin-laden wastes and other chlorinated, TCDD-producing residues.

Or, it may have spread—if spread it had—in far tinier quantities as a regular emission over weeks, months, or years.

Only one thing seemed certain: Nobody in government had done much to find out.

Beyond all the millionths and billionths were the ordinary, child-bearing people of Midland, and they indeed had encountered an unusually high rate of birth anomalies in the early 1970s, just after the big Agent Orange productions. The figures could only be described as jolting ones. In a period that began right around the time EPA announced restrictions on the herbicide 2,4,5-T (because it seemed to be causing miscarriages near sprayed timberland in Oregon), there were fifteen cleft palates among Midlanders when only five would normally be expected.

There had also been thirty-four urogenital defects when only seven should have occurred—nearly five times the expected cases—and heart defects were also exceedingly high.

This was during 1970–1974—just before Carol Jean Kruger began noticing strange things on her farm. Not only did it immediately follow the intensive Agent Orange–making era, but, according to a formal proposal urging a federal study of the birth defects, "it is also known that during the week preceding June 26, 1970, Dow export shipments from Bay City set a new record of 6 million pounds in five ships bound for 'Europe, Latin America and the Far East.' Dow also announced at that time that shipments were 35 per cent ahead of the previous year."

The key man pushing for the study, Charles Poole, now at Harvard University but at one time an EPA epidemiologist and before that with the White House's Council on Environmental Quality, had come across the birth defects in a most interesting way. As part of its argument for the safety of 2,4,5-T in Oregon, when it was trying to stave off the EPA restrictions, Dow had offered statistics from the Midland area to show that seasonal increases in miscarriages were not unique to herbicide-sprayed areas of Oregon.

Put simply, Dow was inferring Midland as a "control," or comparison, group to show that Oregon was not so abnormal. In essence, it was saying: Look, Oregon isn't being affected by the 2,4,5-T, because the same thing happens here in Midland!

Poole had been asked by superiors at EPA to look into the validity of Dow's using Midland for such a comparison, and he began reviewing not only miscarriages but also the birth defects. His eyes widened when he saw how high they were in 1971, 1972, and 1973.

Something seemed to be very wrong, and it carried the ring of dioxin.

At any rate, Poole was not long for EPA. And it wasn't because he was a rabble-rouser. In a calm, unaggressive manner he had begun proposing the study of malformations as a purely academic interest: He intended to take a leave from EPA to complete doctoral work at Harvard, and when he returned to the agency he hoped to begin the research on birth defects as part of his final thesis. The only thing he may have been guilty of was naiveté.

This was during the reign of Anne Gorsuch-Burford at EPA—a front woman in the Reagan administration's drive to ease environmental standards. She was hardly shy about the role she was hired to fulfill, and among staffers her cool detachment from emotionality—and almost certainly from people like those in Midland—apparently earned her the nickname of "Ice Queen."

Burford was the administrator who was forced to resign after allegations of sweetheart deals with polluters, conflicts of interest among assistants, and outright wrongdoing wracked the agency. Though Poole was not the crusading type, according to another EPA official, J. Milton Clark, he was a "highly thoughtful" young man, and the data he sought could have become "a smoking gun in the case for stricter industrial regulation."

It certainly would have been uncomfortable for that big business known as Dow Chemical.

Just before Burford's disgraceful exit from the agency, Poole's academic leave was "denied without explanation" and he was basically forced to choose between furthering his credentials at Harvard and immediately returning to EPA.

41

He was not the type to point fingers, and he had no concrete evidence that his innocent interest in the birth defects was at the root of his sudden dilemma.

But the fact remains that he had little choice but to leave EPA, and the circumstances, he told me, caused him some wonder.

Whatever the motives may have been, the study has never been pursued and the EPA seems intent on preventing another public firestorm like those at Times Beach and Love Canal. On one memorandum concerning the agenda for a meeting about dioxin were these words: "It may be best not to have an agenda because written things do get out."

Besides being careful not to step too far from the Reagan line, the EPA also seemed to be intimidated by Dow Chemical. It knew Dow was used to fighting long and hard.

Never mind spreading VD rumors: Back in 1978 the company had refused the agency permission for an on-site inspection that involved the Clean Air Act, and when the EPA got around that barricade by conducting aerial surveillance with mapping cameras, Dow took the agency all the way to the U.S. Supreme Court in what became known as the "flyover" case. It claimed that such inspections violated the firm's privacy and could threaten trade secrets.

(On May 19, 1986, the Court ruled 5–4 against Dow, the liberal faction siding with the chemical company but the conservatives, led by Warren Burger, ruling that the EPA could in fact take to the air without a search warrant.)

Moreover, according to an internal EPA memorandum dated January 28, 1980, that concerned Hemlock and the disposition of 2,4,5-T and trichlorophenol wastes, "it seems that Dow has chosen to answer some questions in the most superficial manner and some of the most penetrating questions not at all."

The wastes were especially pertinent, since they likely carried thousands of times as much dioxin as the finished herbicidal products.

Though, by 1983, Dow had announced it was abandoning its fight for silvex and 2,4,5-T, it still manufactured 2,4-D (for control of broadleaf weeds), which might carry non-TCDD dioxins; and it still burned the chlorinated residues. At one point, 8.2 million parts per trillion of dioxin had been found in particulates from the rotary kiln. (I saw another figure that put the amount in ash lower, at 280,000 parts per trillion.)

By 1985 Dow and its subsidiaries operated some 440 process plants at 133 locations in thirty-one countries, its total sales more than two and a half times the size of EPA's entire budget. When the EPA attempted to restrict the use of 2,4,5-T, Dow had responded with full fury, piling up impressive scientific studies and legal appeals that would keep 2,4,5-T on the marketplace for more than eight years following this "ban."

Meanwhile, in its defense against Vietnam veterans, Dow's legal documents tallied to three million pages, a mountain of paper that, if stacked in one pile, would have reached halfway up the Empire State Building.

Dow maintained that if the dioxin was under a part per million (meaning 1 million parts per trillion) in the finished herbicide, there was no health risk at hand. Its Agent Orange, said the company, was just 12 percent of a single part. Dow's own carefully conducted tumor research showed no abnormal incidence of rat tumors at 210 parts per trillion, a level quite a bit above a study I mentioned many pages ago.

At EPA's Chicago office, Dr. Clark, a health-effects specialist, picked up the birth-defects ball. He was convinced the elevated rate of cleft palates and other defects that Poole sought to investigate had since been "downplayed" and that "the probability appears to be particularly small that the excursion was due to chance alone."

The "excursions," he suggested, "may be due to fetotoxic or teratogenic agents"—dioxin particularly, he emphasized. He wanted the Centers for Disease Control (CDC) in Atlanta to become involved, along with Harvard University, describing the situation in Midland as a "unique opportunity" to study the possible effects of dioxin on humans—building up a body of data that elsewhere had been so sorely missed.

Such a lack of data, of course, had served as Dow's central argument against dioxin's toxicity.

But the Michigan Department of Public Health gave Dr. Clark little hope it would help with such a study (claiming a lack of manpower and resources), and as of this writing Dr. Clark is still trying to convince CDC to begin a major investigation with or without the health department.

Dr. Clark was haunted too: During 1984, on the way to a meeting in Midland with Diane Hebert and other environmentalists who were petitioning EPA for an investigation, he had stopped his car near a school to ask for directions and, in a coincidence that begged the odds, he saw that one of the kids who came forward had a cleft palate.

I had the same experiences driving the lonely roads on the outskirts of Midland, heading toward Hemlock and the moans of hungry cows.

It may have been a wholly unscientific endeavor—there is nothing a staid epidemiologist despises more than "anecdotal evidence"—but I set out to ask others how their families were faring.

When I stopped to seek my own directions at a modest house in this sparsely populated area, the family rather urgently gathered around me to tell about skin rashes and heart problems that were plaguing their house. And about all the cancer they claimed was tearing through the area.

I heard the same from a young woman standing next to her car. And in the hamlet of Ingersoll was a woman who was at first reluctant to let me in but who once I got inside began pouring forth complaints about fifty chickens that had died all at once in the mid-1970s, and about her dog's leukemia, and about her husband's lupus—a serious immunological disorder.

The fumes hit me as soon as I pulled up the long gravel drive to the McGintys. It smelled like ammonia and chlorine. I was now near the hamlet of Jam, at the intersection of Saginaw, Gratiot, and Midland counties.

Unlike most of those in the area, the McGintys were not farmers. Jim was a college professor, and along with his wife, Marilyn, he had shaped their dream house in a cozy, tasteful fashion, the living room well appointed with Victorian furniture and flawlessly neat, the piano sharply polished.

That smell outside, they told me, was there much of the time. Jim said that when they complained to the state about the odors, someone from Dow would show up before an official would. The employee would be carrying a little tape recorder while he sniffed the air, telling them he didn't smell anything.

"They acted very resentful that anyone would think they're causing a problem," Jim said. His wife added that the state inspectors seemed intimidated by the huge company, whose security guards were certainly a presence along the brine lines that snake through the area.

In a period of just several weeks around the time of my 1986 visit, Dow experienced accidental releases of sulfur trioxide and hydrogen chloride. Also, a worker had been burned when an incinerator flash-backed on him.

Early in May it was reported in the Midland *Daily News* that the county clerk was resigning as public information officer for a Midland council that would supposedly coordinate emergency response to such releases. He was mad at not being told about a code-one alert a short time before. It was enough of an incident to have led officials to keep children sheltered inside a nearby school.

As it turns out, this man in charge of emergency information had learned of the serious release from his *wife,* which led him to believe his position was not being taken very seriously.

Besides the chemicals, Dow also incinerated low levels of radioactive carbon 14—raising the specter of new, unknown interactions with the other compounds.

When I called a chemical operator at Dow, Thomas D. Dauer, he told me in his ten years there he had heard about "quite a few" releases that never had been reported, but they were not intentional ones. He had also heard the rumors about documents being destroyed, and about releases at one time from the trichlorophenol plant.

Now, at home with the McGintys, Marilyn was speaking: "It was in 1976 that we really started thinking something was wrong. We just continued to get ill through 1980. We started having respiratory infections and headaches and the intensity kept increasing. In 1976 our youngest daughter was born and she had constant diarrhea and blood and mucus in her stools. They thought for a while that she had cystic fibrosis. I started having outbreaks like poison ivy. And tachycardia [rapid heartbeat]. The headaches were accompanied by sweats and chills and nausea, cramps, diarrhea. I had allergic shiners under my eyes—it all seemed like immune problems."

Her husband had been given antigen tests and showed two hundred times or so the normal levels, they said. When he approached his doctor with the thought that his family was being poisoned, the doctor "didn't want anything to do with anyone who thought he was contaminated. He thought that was ludicrous. Sometimes I'd go to teach class and my mouth wouldn't work. I had chancre sores on my gums, tongue, inside my cheeks. I couldn't sleep—a pile of amorphous symptoms. I turned gray and had tachycardia too and started wondering about my life insurance."

He dragged himself to school between vicious, all-night bouts of vomiting.

If Jim was having great trouble getting through a class, his wife had the stranger experience of losing chunks of her memory: After eleven years of piano lessons she suddenly forgot how to play!

They told other accounts of a man who would fall off his tractor

45

and go into seizures. And of a stringy, strange-looking bladder a young woman up the road had had taken out. And of a neighbor who'd been in the hospital thirteen times in three years for seizures that were still undiagnosed.

One whole family, they said, suffered from pus sacs in their mouths.

For years Jim and Marilyn took to wearing surgical masks outside their home, hoping this would afford some kind of protection. Like expeditioners in a hostile environment, Jim set about sealing off the house by putting nylon netting in the vents to collect fugitive dust and an electrostatic filter on the air conditioner.

In case the water was also a problem, he bought a distilling unit.

In Marilyn's resigned voice, one last, irresistible dog story:

"We had a golden retriever that Jim shot just last week. He was about four years old when chunks of his fur fell out and he'd get large, open sores and his eyes were nearly swollen shut.

"There'd be points when he'd stagger around and hardly be able to get up, arthritic in nearly every joint, and swollen glands that were nearly the size of grapefruits. When we put him away he had an enormous tumor on his chest."

When I asked if he had noticed anything wrong with the wildlife nearby, Jim said, "I was just out walking the other day and the new dog caught a squirrel, a fox squirrel, and there wasn't a hair on it. It was the damnedest thing."

There were too many other stories about too many other people: more crumbling teeth, more dizziness, more seizures. I remembered one young girl in Hemlock who'd had deformed, black teeth shaped like rabbit ears.

Truly the professor, Jim ruminated about the basic reason for the area's dilemma: Midlanders are proud folks and proud folks were not likely to admit a "tragic flaw" in the very fabric of their community, he said.

The company, after all, had taken a dying lumber town of two thousand and built it into a booming city. In the minds of many, Dow is still thought of fondly.

Mayor Joseph Mann, the chemist, was reached at Dow's electro-chemical subsection. While he did not rule out the possibility of "an unusual component" that was undetectable and had caused an insidious outbreak of disease, the people, he felt, had a natural fear of contamination and it was easy to play on that fear.

"A lot of people who don't have a broad scientific background can easily be made to feel concerned about *any* chemical," he said.

Diane Hebert and her friends, he added, "might be caught up in their own cause. We tried to bring up facts and other studies to allay those fears, but it was as if we were doing nothing. The overall cancer rate is lower than the national average . . . .

"Unfortunately, [they] pick out the thing that advances [their] cause."

The county's health commissioner, Dr. Winnifred Oyen, who had also worked for Dow many years before, decided that the birth defects already had been "watched and screened very carefully" and were in no need of further study.

"If we had a problem we'd certainly want to know about it. With the data we've been following we don't see anything to get alarmed about."

The cleft palates were probably a statistical "fluke," and if there were problems with animals, "the veterinarians would have contacted us."

When I asked about a rumor, passed along to me by an anonymous federal source, to the effect that she had once intervened to make sure the cancer death of a former mayor, Julius Blasy, was not reported as a soft-tissue sarcoma (the final verdict was mesothelioma, usually associated with asbestos), she replied, "I've been in the room when it was discussed, but I was not involved in his actual care," adding: "There was some discussion about it by different people, but I don't remember who said what. I was a listener."

Dr. Oyen recently had criticized environmentalists for causing unfounded fear that the local milk might be contaminated with dioxin. (At the levels tested for, it was not.)

But the dioxin trail was getting hotter. According to recent EPA tests, soil inside Dow's fences had shown dioxin at up to thirty-six thousand parts per trillion (or, as EPA prefers to express it, thirty-six parts per billion).

Such a level can be placed in more meaningful perspective when one realizes that at Love Canal, some surface soil from the most troublesome part of that notorious toxic dump had been six times *less* than the highest level now being tracked at Dow.

47

Dioxin had also been detected in public and *residential* areas near the plant. The positive findings were up to 2.5 miles from the factory.

A smoking gun?

At the least it meant the compound might have touched children playing in the dirt, as well as garden vegetables.

While levels varied—and the EPA, to no one's surprise, decided they were "not an unacceptable risk—some of the readings began to approach the part-per-billion threshold set by CDC. A tenth of a part per billion didn't seem much, but factored at the next range, it was one hundred parts per trillion, or forty times higher than a dose which had caused defects when injected into chick embryos, and six times higher than an Air Force calculation for the level tainting the soil of Vietnam.

While Dow and the EPA were correct in emphasizing that dioxins, in extremely minute quantities, were present in many parts of the country as a result of normal combustion processes (of which we will hear more later), in Midland we were dealing with the most dangerous dioxin isomer, the 2,3,7,8-TCDD, and there was something unnerving in learning that the estimated quantity of dioxin spread about the city, without counting paved areas or building roofs, without counting dioxin that had been there but eventually disassembled under the sun, was 1.1 pounds.

That, in all likelihood, was not as much as had spread through Seveso, Italy, during the world's most serious *known* dioxin calamity. The quantity there was probably between one and eleven pounds.

But, if evenly and orally administered, the current Midland load was enough to kill 300 million to 550 million guinea pigs.

And, indeed, the EPA was finally concluding what so many had guessed and feared all along: "Air emissions from the Dow Chemical plant are the likely source of contamination in the Midland area."

I rode along the Tittabawassee contemplating the very real possibility that this large swath of land had incurred an almost silent dioxin event. If so, Midland and its hinterlands presented scientists with a huge, living laboratory, as EPA's Clark was more or less suggesting.

The cancer rate for the next two decades may contain some further clues, but whatever the final outcome, Midland is a metaphor for many other parts of America in demonstrating the mysteries that arise when technologists lose measure of what they may be dispersing into the wind.

We will encounter other hauntings close to chemical plants, but a loud poltergeist this ghost is not. It is subtle as a constant breeze.

I stopped at a landing where workmen hired by Dow were gathered along the banks preparing for the dedication of a new boat launch the company was donating to the local fishermen just in time for the walleye season. The men were sprucing up the shoreline by cleaning the rocks. A patch of earth was also painted green where sod would not take hold quickly enough.

In other acts of goodwill, Dow had donated $250,000 to the Michigan Department of Public Health for a study of soft-tissue sarcomas, and had set up an educational program to increase public awareness of the need for organ and tissue donations. For such public-mindedness the company had begun receiving health awards.

My last stop was Carol Jean Kruger's farm. She had retired from the role of activist, gladly handing that chore to the able Mrs. Hebert.

But the aging farmwoman gave me two parting anecdotes for officials and Dow scientists to scoff at.

One was the photograph of a Holstein that died mysteriously after rapid weight loss. ("Her liver, when they took it out, it filled a wheelbarrow.")

The second was of a peacock that she insisted was another freak: Despite the fact that it was female, the bird had sprouted a whole plume of fancy male feathers.

Evening came and clouds gathered on the horizon, vaguely threatening a storm. The air showed no smudgy factory plumes, but of course dioxin never was meant to be seen. It was meant only to hound and to haunt; and for all anyone knew, a fraction of it was kiting far and high, moving quietly beyond central Michigan.

6

Miles above the Midwest is a jet stream that sweeps down from the northern corner of the Pacific coast, near Canada, and over the nation's geographical center before moving relentlessly east past the Ohio Valley, the Middle Atlantic states, and out to sea.

This is air moving very fast, and looping like a roller coaster.

At certain times, during a disturbed pattern, the system comes from California. In a reversal of the pattern described above, it shoots north all the way into Canada before dramatically curving down into the states again.

Zigzagging and changing with the time of the year, depending on temperature contrasts, a jet stream, during certain patterns, may course high above the Midland area. From there it speeds over the eastern half of the Great Lakes.

Stirring the air up there, it proceeds to meander around the globe.

A jet stream is a thin, wide swath of air that can move at more than one hundred miles an hour. Often it materializes in a section of the atmosphere known as the tropopause, a sort of gray area between the troposphere, which extends from the ground to a height of about seven miles, and the region of thin air known as the stratosphere, which reaches to a height of thirty-five miles or so.

Sometimes the jet stream splits into two systems, one north, one south. Sometimes it hardly touches America at all, staying up in Canada. During a disturbed pattern it can go just about anywhere, and can therefore transport a particulate to just about anyplace.

The very textbook definition of a jet stream allows that a particle

51

caught in the flow over California could not only find its way north of the American border but could then dip so mightily to the southeast as to show up in Washington, D.C.

Where normal winds might take ten days to carry something from California to New York City, a jet stream, at top speed, could do the job in a single day. Since there is little moisture up there, the swift movement probably would be invisible to the commercial airline traveler. But if enough wetness were available, cirrus clouds might form parallel to the jet stream and indicate its mysterious presence with crystals of streaking ice.

By the time a compound got that high, the amount of dilution would be overwhelming. Most of the time, such a particle would seem to disappear over the ocean, lost to the vastness of air. If it did not fall out long before, any chemicals it carried might be destroyed by the sun's rays, by interaction with water, or by any number of other factors.

But with rain, downdrafts, and simple gravity, a particle can also return to the lower troposphere and wash back to earth intact. It is thus conceivable that an infinitesimal speck of Dow's dioxin could make this long, meandering voyage from the upper Midwest and precipitate onto the steps of the nation's Capitol.

The jet stream, though very powerful, is not really meaningful in the scheme of toxic transport. While dioxin is potent enough in any detectable quantity to cause concern, most chemicals are not much of a worry if they reach the jet stream. Lesser airflows, however, are just as complex and variable, whisking greater chemical quantities across huge territories of America.

According to Kenneth A. Rahn, an atmospheric chemist at the University of Rhode Island, once you get a few thousand feet up you are above more than 60 percent of pollution aerosols.

"The first five thousand feet [about a mile] is probably the most important," says Dr. Rahn. "Most of the material would remain fairly near the surface. Historically many people have a knee-jerk reaction of saying it's the upper atmosphere that transports all the material. That has a historical genesis because studies of transport started with atomic-bomb debris, and in the days of atmospheric testing, most of the debris was injected into the upper troposphere if not the stratosphere—directly.

"With pollution materials the situation is fundamentally different. And we've been very slow to grasp this difference. It's not released with high energy. It's released with very low energy very near the surface.

"It's then put into the more sluggish surface layers, and only some of it ever leaves this layer. There's no problem transporting stuff from the Midwest to the Northeast right near surface layers. That's not to say that it can't transport aloft, because there is transport aloft. But at this point it's very difficult to say how *much* is transported aloft. Measurements are few and far between."

When I asked just how much we knew about the travel of benzoids and other toxic hydrocarbons, Dr. Rahn replied, "This is virgin territory that has really not been looked into by *anybody*. The quantities are often larger than you think."

Most large particulates remain within five or ten miles of their origin, but the smaller they are the farther they fly, and if they are small enough they can remain aloft for days or weeks. At times, if they get caught in the stratosphere, the weeks may turn to months, the months to a year. Gases, if they decay slowly enough, can travel still farther.

The question of atmospheric transport, once keen during atmospheric bomb-testing, has found greatly heightened currency in recent years with the introduction into public consciousness of acid rain.

Unlike the more toxic pollution that we are focusing upon, acid rain revolves, again, around the classic pollution of sulfur dioxide and nitrogen oxide. These are also known as "criteria" pollutants, because unlike most chemicals, these are regulated by specific governmental standards.

Spewed from coal-fired power plants, smelters, and other industries, as well as from petroleum burners such as the average car, sulfur dioxide and nitrogen oxides can become sulfuric acid and nitric acid in the atmosphere once they mix with the moisture there. Their deposition is not just in rain but also in snow, fog, and dew. They also constantly descend in particle or gaseous form during dry weather.

In Europe, Norway has blamed acid rain from England for causing damage to its forests, and trees in the Alpine stretches of Switzerland are also in trouble. A small fraction of the sulfur in Europe even originates in North America.

In the United States, meanwhile, the sulfate deposition in a state such as Vermont comes in about equal shares from regional sources and such relatively faraway places as Illinois, West Virginia, Kentucky, and —yes—Michigan.

The recent specter of lake fish in Canada and upstate New York dying from acid that originates in coal-fired plants along the Ohio Valley and other parts of the industrialized Midwest, as I said, has awakened us to the idea that our atmosphere is not infinite. What you place into it does not just disappear.

In the words of meteorologist and textbook author C. Donald Ahrens of California's Modesto Junior College, the atmosphere "is a thin, gaseous envelope comprised mostly of nitrogen and oxygen, with clouds of condensed water vapor and ice particles. Almost 99 percent of the atmosphere lies within eighteen miles of the surface. In fact, if the earth were to shrink to the size of a beach ball, its inhabitable atmosphere would be thinner than a piece of paper."

Speaking of the release of radioactivity in 1986 from a damaged Russian reactor, Dr. Rahn points to the subsequent tracking of radiation over the United States: "It was only eleven days from Chernobyl to here.

"In terms of transport, this is a quite small planet."

Pollution can fly in the form of tiny droplets, small pieces—or particles—of solid matter, invisible gases, or as a sort of mix of them known as the aerosol.

The particulates may be washed to earth by the falling rain, or fall by simple gravity.

The gases, depending on their solubility, may dissolve into the airborne moisture (moving wherever it moves), permeate vegetation, or condense into fine particulates that can be seen only with an electron microscope.

Many chemicals, including chlorinated pesticides, travel both as passengers on a particle and in a gaseous state. Even metals, under the right conditions, can turn vaporous.

Though it too can have a vapor phase, dioxin prefers to cling to a particulate. It is not water soluble.

On the other hand, solvents such as benzene or toluene, by definition "volatile" compounds that easily evaporate, frequently find themselves in vapor states or dissolved in other substances. Pour alcohol or turpentine onto the floor and watch it disappear with nary a trace. The same is true of many other solvents.

What they are volatizing *into,* of course, is the surface air, which, unlike the jet stream far above, moves in close accordance to local landforms such as hills, mountains, forests, and lakes—thus forging a much more erratic path than the higher, less inhibited patterns. It is not unusual for the surface winds in one part of a state to be moving in a direction opposite those at the other end.

Nor is it at all unusual for these currents to flow opposite to other surface layers. Look up at the clouds and you may see them moving in a direction contrary to that of the wind brushing your face.

If there is a zone of high pressure, where cool air comes down and exerts more weight on the surface (often because it has been piled up by the jet stream), winds tend to diverge or move *outward* from this zone in a *clockwise* direction.

Off the East Coast is a semipermanent system of high pressure, the Bermuda High, that changes position seasonally and sends air moving in the clockwise fashion through the Southeast to the Mississippi Valley and then northward and back toward the east.

It is a little like standing in the Atlantic, aiming toward Florida, and throwing out a boomerang.

A low-pressure area, on the other hand, causes a *counterclockwise* flow, sucking air toward itself like a semi-vacuum.

I am speaking in simplifications and generalizations that might madden any number of chemists and meteorologists, but the alternative, because of the complexity of air patterns and the multivarious character of many pollutants, would be the even more maddening scenario of convoluted detail.

The Midwest, to say it simply, is a real potpourri of such wind factors, and as such it can be seen not only as the nation's heartland but also its lungs. Though it can receive substantial flows from an easterly direction, it seems mostly to inhale from the north, west, and south, exhaling towards the east.

It could also be described as that vat in which ingredients are constantly poured in and the finished stew, again, is constantly burbling out. Indeed, the nation's major wind patterns converge over the central states and cause swirls on a meteorologist's map that put one in mind of a caldron.

From the Rockies comes what we could call the Great Westerly Flux. Throughout the Northern Hemisphere, because of the deflective effect of the earth's rotation, the general pattern is for the wind, especially at high altitudes, to flow eastward.

A westerly is wind from the west.

But along the way other forces come into being. One of them is a polar front of air that presses into the pattern and could be called the Canadian Cool. Simply, air from Canada.

Its counterpart is the warm mass from the Gulf of Mexico that evolves in great part because of the Bermuda High—a strong, highly important movement of air that often moves north and then, at the upper Midwest, is shaped by pressure factors and the relentless westerlies into an eastbound curl.

This "curl," to put it simply, is like the tail of a whip.

The pattern is greatly complicated by other systems of high and low surface pressure that, from day to day, pop up in widely different spots on the mainland United States. For example, there might be a "high" over Colorado and another over North Carolina, with a "low" up near Canada.

Or maybe a low in Nevada and a high in Georgia.

Though meteorologists swear there is a pattern to these, it seems to the rest of us like shaking dice in a tumbler and throwing them on a map.

But for the sake of visualization let us pretend the flow from the Gulf is simply a northeastward curvature. Because the curve tends to gradually slash back and forth, turning east at widely varying points not only in the Midwest but also the South (and as such resembling a river that heads in a general direction while often adjusting the details of its course), we can call this the Gulf Superwhip—the tail slowly snapping up and down and the stem moving east and west, back and forth.

It was the Superwhip that carried up the toxaphene.

There are so many different little wind patterns on any given day, at all the various altitudes and ground points, that mapping them all would be comparable to dumping a bucket of needles on the floor and factoring in the direction of each.

This is another area where technology is not quite up to the task. It has only been sixty years or so since the concept of weather fronts was first articulated, and while supercomputers now provide dazzling three-dimensional models of air patterns, it's impossible to present a full and accurate picture of them on a single map, let alone in a single chapter.

For one, the characterization of such atmospheric movements may differ according to the number of years' worth of observation a meteorologist is averaging in. A pattern based on five years of localized wind activity might look somewhat different than one based on ten.

We are not even certain all the time precisely where the jet streams flow.

But there certainly are the basic trends, and one useful tool for presenting them at the surface level is a device known as the "wind rose." It is basically a chart with a circle in the middle. The circle represents a particular geographical vicinity, and from it branch sixteen graph-bars, each of varying length. They symbolize the percentage of time wind blows from a certain direction.

And, as such, the wind rose also shows in which direction a locality is spreading its toxics.

It got its flowery name because, in the strangely geometric minds of engineers, the bar-graphs extending from the center somehow resemble rose petals. To the average onlooker, however, the graph would look more like a broken wagon wheel—with long and short spokes. If the wind comes from the west most of the time, the spokes on the western part of the wheel would be the longest ones, and the rose is therefore lopsided, the petals (or spokes) on the eastern side—where little wind is coming from—looking stunted or broken.

The region receiving pollution from the center of the wind rose is the region on the short end of the directional stick.

Where the annual wind roses from a place like California tend to be grossly lopsided to the left, representing the dominance of westerly currents, at the central region beginning near the Mississippi River they become more symmetrical, the prevailing westerlies still prevailing but the winds by now coming more evenly from all sixteen compass points.

In a place such as central Michigan, the predominant wind comes directly from the west 11 percent of the time. That is twice as much as comes from directly east. The most predominant flow, at more than 12 percent, is from straight south—from the stem of what I took the liberty of nicknaming the Superwhip—while more than 40 percent of the contributions are made by points in between. That is, from the northwest and southwest.

Central Michigan, in short, is more likely to get its air from Wisconsin than it is from Pennsylvania. Pennsylvania, and everyone else east, is on the receiving end.

Where the airport in Flint, Michigan, might record an average annual input from a northeast point only 3 percent of the time, the input from a point on the southwest side of the wind rose might record flows from that direction 10 percent of the time.

In other words, the spokes on the left side of the wheel are longer than those on the right side and demonstrate that the air from a place like Midland moves strongly east, toward the nation's largest population center.

As important as the wind direction, of course, is the nature of the chemicals themselves. "Compounds vary as far as their reactivity," says another meteorologist with experience tracking pollution, Nathan Reiss of Rutgers University in New Jersey. "Some of them travel for thousands of miles, some will deteriorate quite rapidly. The atmosphere is a chemical cocktail, and our measurements are very crude at this point."

A compound like benzene deteriorates quite rapidly in the atmosphere, oxidized by bacteria or chemical reactants that can bust up the ring and turn it into more of a chain. Its atmospheric half-life (the time it takes for half of it to break down) is, under normal circumstances, less than a full day.

But with winds of ten miles per hour, Dr. Reiss notes, the compound could travel 240 miles before burning out.

There is certainly enough benzene around to ensure that some survives. About 28 billion pounds of it are produced worldwide each year, and it can generally be found in ambient air at levels of one hundred to fifty thousand parts per trillion, even in remote regions.

This should come as little surprise, considering that its prodigious uses have included that as an important intermediate in the manufacture of many other chemicals.

Also, it must be remembered, benzene exists naturally in petroleum.

Dioxin can also break down when exposed to the elements. Sunlight is sometimes enough to do the job. But no one is really sure how long it lasts. Experiments have yielded conflicting results. The compound loves to grasp onto ash and other particles, and estimates of its half-life, in soil, often mention a year.

Irradiating it on a glass surface with sunlamps, some experimenters have reported dioxin's half-life to be only 5.8 days. On the other hand, at the Love Canal, where it was quite effectively shielded from sunlight, and where there was no one beaming it with sunlamps, dioxin was present in frightening quantities nearly thirty years after it had been dumped there.

In the air, other compounds, such as carbon tetrachloride, seem to last nearly indefinitely.

Which brings us back to what is perhaps the single most significant category of air contaminants: those "VOCs," the volatile organic solvents, which so readily take to the wind.

Though they find many uses besides as degreasers, I call them "solvents" to distinguish between them and such chemical categories as metals and pesticides (which, to keep matters complex, often overlap each other).

We could also categorize chemicals as to whether or not they are chlorinated, and whether they are organic (like benzene) or inorganic (as metals are).

But to stay for now with the first method, another category might be called "POCs," or products of combustion. Here dioxin would be found.

While the world of modern chemistry is much too convoluted for such ready categorization—solvents are used to make ink, rubber, and pesticides, as just a few examples—it is still helpful for our immediate purposes to try and reduce it to a few encompassing species (just as it was helpful to see general patterns in a million weather fragments).

So, "solvents" is how we will think of a wide array of airborne compounds. As in Midland, benzene, toluene, and methylene chloride will appear everywhere else—more ubiquitously, perhaps, than any other class of compounds.

Other common solvent-type materials include chloroform and carbon tetrachloride, two substances found to act as carcinogens in laboratory animals. The compound xylene, which finds use in photographic solutions and like many solvents can cause liver and kidney damage, is detected, along with toluene, in ambient air near car factories where paint is sprayed.

Though not finding its major use as a solvent, another volatile organic, formaldehyde, is also all around the place. It is encountered by every citizen of the land on a daily basis. Besides being another compound that comes out of car exhausts, formaldehyde has been widely used as a fumigant, a disinfectant, in cosmetics, in papermaking, in home furnishings, and in embalming fluid. It was also used in foam insulation.

Formaldehyde's points of attack are the eyes, skin, and respiratory system. Some relatively innocuous chemicals change to formaldehyde as they react in the atmosphere. Once ingested, wood alcohol may react *inside* the body to form formaldehyde.

59

For perspective, however, we must keep in mind that while, in great enough quantities, such compounds can have the most dire of consequences, including chromosomal or other genetic-type aberrations, dioxin of the TCDD isomer has caused government hand-wringing at levels in the neighborhood of a billion times lower than the level of official concern for benzene or formaldehyde.

Where these two compounds are considered a problem above a few parts per million, dioxin causes some scrambling about when it is found airborne in parts per *quadrillion.*

Like a large number of other chlorinated chemicals, it may stay for a long time in the body.

On the other hand, many of the simpler solvents and other volatile compounds tend to leave the body rapidly. They are often quite soluble and thus amenable to excretion through urine and perspiration. Or, they are readily metabolized into soluble materials. Because they tend to evaporate very rapidly, they also volatize from the blood and can be exhaled.

The hydrocarbons and other organics are called "organic" because they are built with carbon—the building block of life itself. While carbon-based chemicals are hardly living organisms, living things are also founded upon it. Carbon is the basis for molecules of protein, carbohydrates, fats, and enzymes.

All told, about 20 percent of the human body is made of carbon. It easily combines with other carbons or atoms of different substances. There is just about no end to the configurations that carbon can form. The diversity ranges from a blue whale weighing more than a hundred tons to microscopic bacteria.

The simplest organic compound made of hydrogen and carbon is methane, a gas of decomposition that can rise from the local garbage dump or from decaying swamplands (causing UFO reports when it ignites in the air).

By itself, methane is basically harmless: a single carbon atom surrounded by four hydrogens.

Replace one of the hydrogens with chlorine, however, and you have methyl chloride, a solvent the body would not so naturally encounter. Replace two and you have methylene chloride, the paint remover. Replace three hydrogens with chlorine and you have the solvent chloroform. Replace all four and you have carbon tetrachloride, which rivals benzene in toxicity.

Make a configuration that looks like two benzenes locking bumpers and you have napthalene. Or put a little bridge between them and a biphenyl is formed.

Take out the hydrogens in the benzene ring and replace them with chlorines and you have an old, persistent insecticide that, as a chlorinated hydrocarbon, is in the same class as DDT—benzene hexachloride, or "lindane."

The isomer of dioxin we are most concerned about is also known, we should recall, as TCDD. The abbreviation starts right away in telling us about itself: tetra-chloro-di-benzo, meaning four chlorines and two benzoid rings.

As one more frame of reference, before the action gets hot and heavy, we have to delve into the domain of spookery again. This time it is a set of compounds that the average person is not familiar with yet.

NEW TOXIC THREAT IMPERILS AMERICA; DANGERS SEEN FOR EVERY PART OF NATION; SUBSTANCE IS NEARLY AS POTENT AS DIOXIN AND MORE WIDE-SPREAD

None of America's sedate newspapers have jumped to any such headline, and neither have the city tabloids. By and large they simply have not been given the full scoop yet. And if they had, a grumbling editor may have nixed the story anyhow, arguing that this "new" contaminant does not fit in with the chemical lexicon—the DDTs, DBCPs, and PCBs—to which the media (and therefore the average person) has been programmed to respond.

But this new set of compounds, when carried by the wind, is a true alarm-ringer. It is far and away more toxic than any insecticides or solvents we normally concern ourselves with. Traveling on particles of ash, it is as much of a public health threat, in fact, as the worst type of dioxin.

The set of compounds is known as "furans." Its technical name is tetra-chloro-dibenzo-furans, which right away reveals its benzoid status.

It is also another product of incomplete combustion.

The structure of furans, we should mention briefly, is astonishingly similar to TCDD. The isomeric names for the most poisonous of both sets are basically identical but for the last half a dozen letters. Where the shortened version for the most toxic form of dioxin is 2,3,7,8-TCDD, the shortcut for the most deadly furan is 2,3,7,8-TCDF.

Indeed, an untrained eye would swear it is seeing a dioxin when really a furan is involved. The only difference is that the two benzoid rings in the center of furans are connected with only one oxygen atom, while dioxin is stitched together by two.

More importantly, furans (we will use this to speak chiefly of the TCDF isomer) are only two to ten times less toxic than TCDD.

In the air it is much more dangerous than uranium.

When one considers what we said about dioxin being millions or billions of times more toxic than poisons like benzene, a tenfold difference in toxicity is a mere drop in the bucket. Furans, like a true phantom, can trigger biological responses at virtually undetectable levels.

Worse, as our headline mentioned, this compound appears to be much more pervasive than dioxin. Any number of materials, when they are heated, can create it.

"There is extremely high exposure—from the incinerators, from heated PCBs, from your automobiles using leaded gasoline, and at the same time we know that structurally [furans] are very stable in the environment, as stable as 2,3,7,8-TCDD," says Dr. Debdas Mukerjee, a molecular biologist and senior science advisor for the Environmental Assessment Office at EPA. "And TCDD is an *extremely* toxic chemical. When you compare furans with that and say that it is three or ten times less toxic, it also is a *very potent one!*"

In theory, all it takes to form a dangerous furan is the burning at the right conditions of compounds carrying the correct array of chlorine, hydrogen, carbon, and oxygen. (The same is true, to some extent, of dioxins.)

This means that furans can show up in chlorophenol processes, in wood-waste boilers, in plants handling chlorides of metal, in hazardous-waste incinerators, in copper and steel manufacturing, in such heavily used industrial ingredients as hexachlorobenzene, and, to a lesser extent, in the exhaust of cars after a chlorinated fuel additive has been heated.

Furans will also be found virtually anywhere there are polychlorinated biphenyls, the PCBs. PCBs were chiefly used as coolants in electronic insulating fluids and still exist in many transformers and compressors.

Of all the potential sources of furans, however, the one looming most ominously is probably the large municipal trash incinerator. Because of the discarded plastic and other halogen-carrying compounds, modern garbage provides all the raw materials a furan needs in order to form.

Concern in the academic community has been growing steadily

since the 1970s and is reaching a quiet crescendo: Scientists are perplexed about the lack of public attention to this "new," virulent, and apparently ubiquitous compound.

"There's no question that we have to consider the furans and dioxins together," said Alexander B. Morrison, former assistant deputy minister of Canada's Health Protection Branch, at a symposium on the subject in Michigan.

"How to do that, how to introduce the concept to the public, is extremely difficult without losing all credibility in the process. The public has been so frightened by the dioxin issue that it is hard to say that not only are there *those* terrible things, but there's a whole class of *other* terrible things and you're not just going to be killed *ten* times, you're going to be killed *fifty* times" (my emphasis).

Furans like to kite around. If a particulate gets high enough, it might find a ride on a jet stream.

But it would be impossible to gauge their human results once they become so scattered. To get a real look at the effect of airborne chemicals, one has to move, again, closer to the emission points.

And the closer you get, the more it seems like a tragedy.

7

"We have babies that had their brains outside their heads," said Hazel Johnson. "From what I can understand, the one was aborted at seven months pregnancy because of it. You know how they take those ultrasounds? And you could *see* it.

"The second case was a little girl; she was really tiny, extra small, you could have put her in a shoe box, and her eyes popped out. They were saying you could see the brains outside the baby's head. It was the talk of the neighborhood. It lived to be three months old, right there around 131st Street.

"And next door to them they had a three-year-old girl, died of a cancer. She was taken back and forth to the hospital since she was *born*. And right here in back of me is a little girl who has a hole in the top of her nose, in between the eyes, and the bottom of the nose is smashed in and meets with the lip, and one eye is partial-closed and the other to the side, they told me. She looks so bad nobody wants to look at her."

Mrs. Johnson fidgeted and exhaled thickly. Her disgust, I realized, was a cover for her fear. Lately the smells had been so strong that one of her friends passed out because of it, she said. And the kids had watery eyes, which made her all the angrier. She is a maternal black woman with four grandchildren and seven children of her own, and so her apartment was heaped with baby food and soda pop. Handy atop the refrigerator was also a can of roach spray, for this is deepest Chicago, the housing projects on the notorious South Side.

In came a wide-eyed little girl looking to her buggy but finding the adults too busy to take her out. Their thoughts were on the youngsters who never made it. "And then there's another case of a baby that died after two years of cancer. That was a little boy. And a baby that was two or three months old and the mother was telling me that the baby was dead two *years* before they actually found out the cause of the baby's death, and they were saying it comes from some kind of paint solvent that was in the baby's system. There was a baby born with no eyes or mouth, and one born with no sex organs."

Hazel's own home sounded like a wheezing, rasping hospital ward. She suffers from angina, two of her daughters have been losing hair, a grandchild has asthma, and one of her sons has also had difficulty breathing. So did her husband. He died of lung cancer in 1969.

She came close to having four other grandchildren—in fact, they were two sets of twins. But one set was lost in a miscarriage and the second set was stillborn, succumbing, Hazel told me, to something the doctors called "toxemia."

If what the residents described is even partly accurate, there is tragic evidence here that could help put to rest any remaining doubts about the potential of our modern landscape to shorten, degrade, and otherwise infringe upon our lives.

The sky is red, as if the whole ghetto is one big blast furnace, and within a short distance of Mrs. Johnson's are chemical plants, illegal dumps, massive steel mills, heaps of rotting city garbage, wastewater sludge that is exposed to the air, a chemical-waste incinerator so special it is known throughout the nation, and landfills that accept the region's most hazardous wastes.

In fact there are thirty-one landfills around these parts, and twenty-two factories that are officially described as "major point sources" because each releases one hundred tons or more of at least one kind of pollutant annually.

The wind seldom brings much relief to the projects, only more fumes and fly ash and doubtlessly some furans.

There have been so many frightful signals here that citizens have taken it upon themselves to conduct their own health surveys. Hazel Johnson has done two of them. The first elicited a thousand responses, she said; the second was still in progress. "We been having that smell so strong that I don't want to go out my door. It made us all sick, nausea, like we *all* wanted to pass out."

Martha Kindred, a board member at the neighborhood health clinic who has both asthma and a congestive heart problem, was also there, along with a young man named John Cherry, who had his own accounts of cancer ravaging people who seemed too young. He has a son born with a hole in his heart, he added, and when Hazel mentioned skin rashes in the vicinity, he said, "You talking about discoloration of the skin? I know I been having that—me and my brother, my stepfather. I get it in blotches, a lighter-color brown."

There was also talk of greenish clouds, silvery fallout, and children with nosebleeds. It comes with the turf. In the annals of American environmental history, Chicago occupies an early and important niche. When the city was incorporated, in the 1830s, one of the first acts of the municipal government was to pass an ordinance against the disposal of dead animals in the Chicago River. The effects of biological pollution had been awesome ones. By 1854, according to a state report, one out of every eighteen residents died of cholera, and there was also an epidemic of malaria, which is literally translated as "bad air." In the miasma of swampy air, germs were spread about by the flitting, flying mosquito.

Aware that acute problems were also related to man-made smoke, the city developed standards for ventilation in the mills and other workplaces as early as 1910. It was initially a wet, desolate, reedy area with flocks of geese and grazing sheep.

But Lake Calumet and a river by the same name were there, providing the raw requisites for a harbor and canals, and starting in the middle of the nineteenth century the iron and steel companies had begun flocking there like the geese. It was also the area in which George M. Pullman began his Palace Car Company.

What had begun as a community of farms and rendering plants was soon turned into an industrial thicket comparable to the Midlands of England or the Ruhr Valley in Germany. While Mrs. Johnson's neighborhood is known as Altgeld Gardens—a peculiarly flowery way of describing such a downtrodden and squalid place—the names of two other area neighborhoods are more to the historical point: "Irondale" and "Slag Valley."

The very name of the region itself—"Calumet"—comes from the Indian term for the peace pipe, with which smoke was offered to the demons.

By 1960 residents were still so upset with fallout from the steel mills that they had marched on the office of Mayor Richard Daley. That was a full ten years before Earth Day. Now, more than two decades hence, the hellish smoke was less visible but perhaps more dangerous—the ephemeral vapors of benzene, xylene, and toluene.

In March of 1986 the Illinois Environmental Protection Agency released a report showing that these chemicals were in the air of southern Chicago along with the metallic elements arsenic, beryllium, chromium, and cadmium. With the exception of xylene, all are listed by the U.S. Department of Health and Human Services as carcinogens.

Used as a protective coating for steel, cadmium, as an escapee into air or water, can also lead to severe kidney damage and an extremely disabling type of emphysema. Though it was found in one soil sample at 13.2 parts per million (where .01 to .7 parts per million would be the natural range), the state concluded that such pollution did not indicate an acute problem.

"However," the report warned, "we cannot be quite so certain about the more subtle, long-term adverse impacts which may well be taking place, but which lie beyond our current ability to fully document. The ability to be more conclusive about the effects of lengthy exposures to very low levels of chemical substances requires a new set of regulatory tools."

As I leafed through Mrs. Johnson's health questionnaires, which probably contain as many "long-term adverse impacts" as most people would ever want to see, another neighbor named Cleonia came in, describing the skin rashes and loss of hair in *her* house, the rashes looking like "you got the hives or something." She also had had a stillborn infant, she said. "At night you could be sleeping and the air is so bad it wakes you up!"

On the survey forms another resident wrote, "You have to cover your face to breathe."

The air was "never sweet!" wrote yet other complainants. It smelled like "something dead!" A young man complained that this atmosphere had "affected my sex life very seriously."

Inversions occur just after sunset, when the wind dies down, and the only relief comes from cool northeasters.

Many of the people said the odors lingered all day every day, especially after a storm.

As for the unending stream of health complaints, which seemed to spare very few homes, there also appeared to be an excess of swollen limbs and thyroid problems. And Hazel warned that the people probably didn't list all their ailments. In some cases they had been too embarrassed to admit to cancer, forcing her to pry such information out of them. "We've got a lot of kids that is retarded, and the people keep them in the closet, too."

In other cases, however, the problem of accurately reporting their symptoms seemed to be one of education. Most of the afflicted could not properly spell the diseases they had ("canser" in one case, or malignancies in the "lump nodes").

Another began the questionnaire by describing her family's problem as a case of eye irritations, but then added, "My daughter has cancer and had to have her leg removed at age 9."

I called Valerie Kyles, a clerk-typist and one of those who Hazel said had a birth-deformed child. She very capably explained that, in reality, the problem with her baby was not that the brain had been outside its skull but that it didn't have much of a brain to begin with.

"It was anencephaly," Miss Kyles recalled. "Means the baby's brain didn't develop all the way. It didn't have the brain, or it was an undeveloped brain. They kept taking ultrasounds and they couldn't get a good picture on the baby.

"Well, most of a baby's weight is in its head and brain, and so they were saying 'Why can't we get this baby?' And then they called in a specialist and he told them there was no brain in the head and the baby was always floating around in me. They said it would have died anyway, so they took that baby. She'd have had a real small head.

"And I've had miscarriages—two. One time they suspected me of having cervical cancer, and even now, my son, if the chemical count is high, his eyes turn *so* red, you know, and I used to say he was getting a cold.

"But that's not it. They turn *blood red.*

"When I was pregnant with this baby they took a Pap smear and it kept coming back 'class two,' so that meant there was something wrong. That went on for the nine months I was pregnant, and then, nine months after that. I also had a fibroid.

"If they release those chemicals, you can't breathe," she added. "Or your sinuses go haywire."

The birth deformity that her one child suffered, anencephaly, isn't rare by any means, so it would be difficult to pin on the pollution. And Hazel Johnson's somewhat askew description of it hints at the tenuity of hearsay accounts.

However innocently, people tend to mix up details and use hyperbole to get their problems heard.

Whether in the ghetto or an affluent suburb, a piece of medical gossip can quickly turn into a full-fledged rumor, the rumor then heaped with enough exaggeration to turn it into a legend, the legend soon expanded into the authoritative tale of a frightening epidemic, and, finally, the epidemic—in need of a point of origin—is linked to the most obvious and therefore the most convenient source, whether that be an unsightly smokestack or a smelly landfill.

To compound matters, ghettos like that in southeast Chicago long have been notorious for spawning birth problems through causes other than toxic chemicals. Drug use, malnutrition, and generally poor prenatal care are often cited as such in urban studies. Germ-carrying rodents also tend to abound, along with mangy, straggling dogs that pick through litter in the fields and alleyways.

Other ailments can be caused by psychosomatic reactions. People feel sick when they believe they *should* be feeling sick. The very stress of hearing about a toxic problem can lead to shaky nerves and splitting headaches, which, at the same time, can also be very real, clinical manifestations of an actual toxic exposure.

Picking between those and a myriad of other possible causes is at best a difficult chore. Once an environmental problem *is* established, too often every subsequent ache and sniffle is related to it—a sort of chemical McCarthyism. As we encounter ever-greater horror stories, we must remember that environmental effects are often in the eye of the beholder.

But Mrs. Johnson and her neighbors were not imagining the bogeyman. When the Illinois health department looked at cancer mortality in certain community areas of south Chicago, it found that there were indeed excess deaths from the disease, with lung cancer in white males, bladder malignancies in white females, and prostate cancer in a group of elderly white males significantly higher in some cases than the rest of the city.

That was saying a mouthful. Cook County, in its own right, had a rate of cancer more than 10 percent above the national average, I noticed. From 1970 to 1979 the rate for white females in Cook was a full 17 percent higher than the rate in Midland, Michigan. A review of cancer mortality by the Illinois Cancer Council also found certain clusters or elevations of esophagus, pancreatic, and cervical cancers in the western and southern parts of town.

Clusters of leukemia appeared in both the far north and southwest of the Windy City.

But quite predictably, the state and other researchers steered clear of directly blaming the environment, mentioning that the origins of such high rates probably had as much to do with smoking, ethnic origins, and occupational exposure as anything else. As far as the ethnic origins, the council's study pointed out that leukemia often tends to be high among Jewish people, while the same is true of cervical cancer among Hispanics.

It is always easy to blame cancer on cigarettes and other excesses of life-style, but it is not always the whole truth and nothing but the truth. Many times it seems to be too convenient an explanation.

Credibly determining the extent and origins of disorders in a neighborhood like Altgeld Gardens calls for on-the-scene investigation by a trained medical detective, the epidemiologist. Ideally, residents should be interviewed in person, their medical records pulled from confidential files, the cancers checked to make sure what is being called a "brain cancer" did not actually originate in another organ and then spread there.

In the most reliable epidemiological studies, samples of blood are taken from the people and checked for enzyme levels. This might reveal abnormalities in the liver—an organ exposed to a wide diversity of chemicals since its very job is to help break them down.

The problem is that, like other physicians, epidemiologists are not fond of making house calls. And in environmental trouble-spots, door-to-door evaluations can make or break a case.

Such an investigation is the only truly thorough way of judging both minor and major symptoms in their proper context.

It seems most studies are instead based only upon existing health data. This often involves a review of death certificates which is thus known as a mortality analysis.

That means it is basically a retrospective study.

The problem with studying a disease such as cancer in retrospect is that cancer can take ten to thirty years for development, meaning that an epidemiologist looking at mortality statistics is not necessarily looking at an ongoing trend but at exposures that took place (or didn't take place) many years before.

Moreover, such evaluations miss those residents who were exposed in a region like Calumet but who moved away from the area before discovering their illnesses.

"I think the statistical methods and tools have been useful, but I think they do miss a lot of things," said Dr. John Noak, chief of the Illinois department of health information and evaluation. "And of course as to what the cause is, you really need field work, or 'shoe work.' Statistics are at best one way of describing reality."

When a clear, ongoing excess of disease *is* apparent, of course, that

71

hardly means that ambient air is to blame. There could indeed be a preponderance of heavy tobacco smoking in the study group, or the people may be exposed to hazardous substances at work or from household products that have nothing whatsoever to do with the prevailing winds.

A runny sinus can result from nail polish or Mrs. Johnson's roach spray.

As far as proving cause and effect, this points up the greatest barrier of all: the fact that there are nearly always so many variables in a given person's environment, so many different types of exposure, so many factors that might relate to cancer, that no conclusions can be drawn.

The situation in the south of Chicago seemed to transcend such scientific hedging, however. As Lee Botts, at one time an employee at the federal EPA's Chicago office and former chairman of the Great Lakes Basin Commission, said in criticism of the state's environmental report, "I don't have to go through all this to be convinced that that's not a fit place for human beings to live."

Indeed, the manifestations were the type that are apparent to just about everyone except industrialists, governmental regulators, and the rigid, deskbound, protocol-driven epidemiologist.

Seldom does an epidemiologist even acknowledge an unusually high rate of illness in a given neighborhood, let alone link such an illness to the environment.

And virtually never will he specify a single chemical as the cause of human death unless it is a case in which a blatantly visible accident has caused an immediate body count.

8

I headed from the Chicago slum toward the outlying industrial areas that extend well past the Illinois border and into Indiana. Along the way I stopped in front of the hazardous-waste incinerator not far from Hazel's house. It is now owned by Waste Management but at one time was the property of SCA Services, a Boston-based company that had been investigated for ties to the Mafia.

Waste Management itself has not been linked to organized crime, but it is certainly a corportation that has had problems with civil law. In one instance it drew what officials in Ohio believed at the time to be the largest such fine in history for practices at a waste site in Vickery.

The site's former head chemist accused the company of illegal waste dumping and record tampering, charges Waste Management says were nonsense. More recently Ohio had fined the firm for the release of an acidic pale-yellow cloud from a holding pond there. The conglomerate was also subject to several lawsuits filed by the Illinois attorney general.

In Chicago its incinerator, which can burn 8,922 pounds per hour (and runs twenty-four hours a day), represents one of only three major commercial burners in the country that is permitted to burn PCBs, the biphenyl electrical coolant which has caused contamination—and controversy—in dozens of communities.

When PCBs are set to burning, the question of furans and dioxins naturally crops up. "If you've got a PCB molecule put into an incinerator, you've got most of the furan structure there already," I had been told by Ronald Hites, professor of environmental science at Indiana University in Bloomington. "All you've got to do is add a hydroxy group or an oxygen on an 'ortho' position to close the rings, the position next to

73

where the rings are connected. So you really are starting with a structural feature that is nearly the entire furan molecule to start with."

But furans can be destroyed at the high temperatures this incinerator is capable of reaching, experts claim, and it is difficult to think of such a terrible poison when looking up at the majestic white plume tumbling out of the stack.

The incinerator itself is painted a spotless white and shaped thick and vessel-like—as modernistic and good-looking as a trash burner can be.

Whiffs of steam quickly dissipate as if being harmlessly teleported out of our very dimension (and for all we know, harmless they are).

But before it disappears, the plume rolls upward like the climactic clouds in a Steven Spielberg movie.

The other landscape is not nearly as pretty. Aside from the mountainous landfills littered with paper and gulls, there is also PMC Specialties Group, a facility where ink pigments and other chemicals are manufactured. It had once been owned by the Sherwin-Williams Company, which continues to operate other units on the site that make resins and coatings for industrial consumers. Exactly where it came from I couldn't tell, but in front of the large industrial complex was a very strong, caustic odor that soon set me sneezing and closing the car window.

On April 24, 1986, the Chicago Lung Association and another public-interest group, Citizens for a Better Environment, released a report in which they attempted to gauge what kinds of material were getting into Illinois' air. They were not interested in the traditional particulates, sulfur or carbon monoxide, but rather the same toxic, noncriteria emissions we have been hounding.

"The potential for environmental exposure to these hazardous air pollutants has increased significantly over the last three decades, primarily because of the dramatic increase in our use of chemicals," the report explained. "Since World War II, chemical production has increased tenfold, from less than 20 billion to over 220 billion pounds per year."

But when they set out to cull, decipher, and tabulate state-owned data, the study's authors were met time and again by very sketchy, piecemeal information and long delays. Translation: The state of Illinois, like the vast majority of states, was not certain *what* all was getting into the air and was in no hurry to let everyone know that.

Environmental bureaucrats, in fact, had collected figures on only a fraction of the potential noncriteria sources, and whole factories and manufacturing units were missing from this data. The two inquiring groups had to wait nine months for permission to review the permit files of six companies in which they were especially interested because, under state law and regulations, a company has the right to declare process and raw-material information "proprietary information"—a trade secret.

When finally they were able to dig into government files (and also into a toxic-emission inventory that was not secret but contained such limited information it all but guaranteed any figure they could extract from it would be a significant underestimation), the investigators calculated that throughout Illinois, 14 million pounds of the fifteen most common hazardous pollutants are emitted each year.

A quarter of them pour out of Cook County.

Often the vapors move across downtown Chicago and over Lake Michigan, hovering in a low haze. Or they drift into the carpet of lake clouds.

There they are cooked into ozone and swept by a breeze back to shore, continuing up to Wisconsin, Michigan, and Canada—or east to Ohio and Pennsylvania.

In one study funded by the General Motors Corporation, scientists using both ground-level and airborne samplers found that near Janesville, Wisconsin, which is about eighty-four miles northwest of Chicago, the air was carrying a wide array of solventlike materials—in fact, at least forty of them, including toluene, xylene, trimethylbenzene, ethylene, butane, and methylchloroform.

The solvents were growing longer in name as chemists continued to combine, reorganize, and generally play with the molecules. As for short and snappy xylene, it can attack nerve cells and lead to cerebral dysfunction.

Moving over the border to Indiana brings refineries in sight. Just before the state line is also an abandoned brewery.

In every direction is steel. The south shore of Lake Michigan is still fairly sinking under the weight: Inland, USX, and LTV.

I was heading for East Chicago, Indiana, which is to Chicago what Jersey City or any number of other towns across the Hudson are to New York City.

In other words, this was not the scenic route.

But the setting sun was playing off some of the factory smoke and turning it into a flowing prism of violet and blue. A layer of red and rather dark gray smog formed a somber backdrop, telling the visitor immediately that there was a problem with particulates, nitrates, and hydrocarbons.

I was in the company of three local activists who believed, like Hazel Johnson, that their families were endangered by the pigmented atmosphere. They were Lydia Meyer, a schoolteacher; Colleen Aguirre, a hairdresser; and Marge Surufka, a housewife and part-time bookkeeper. They were currently fighting various plans to burn hazardous chemicals in their community, and they planned to hold a mock funeral procession soon, complete with casket, pallbearers, and a jazz band to express their fears.

The march would take them from the site of one proposed incinerator, at a Stauffer Chemical plant in adjacent Hammond, to the house of the mayor there, Thomas McDermott (who, because they viewed him as supporting Stauffer, had been nicknamed "Toxic Tom"). The women also planned to mark off a 2.2-mile radius around the plant with morbid black ribbons.

At one demonstration they'd already dressed in masks and moonsuits, releasing black balloons attached with mail-back cards in front of the plant. The idea was to see just how far the air from there traveled.

One of the cards was mailed back from South Point, Ohio, about 330 miles to the southeast.

Stauffer's plant manager, R. D. Miles, was not amused. He called such efforts "a fear campaign" based upon "rumors and half-truths."

On September 21, 1986, the firm struck back at the protesters by taking out a full-page newspaper advertisement with a photograph of the plants' ninety-eight employees under a huge, bold headline that said, "WE LIVE AND WORK HERE TOO!"

Stauffer pointed out that incineration of wastes prevented such chemicals from seeping out of dumps and into water supplies. It also said its method "is 500 times better than EPA requirements."

Another part added, "Unfortunately, during recent weeks a skillful campaign of misinformation directed largely against our incineration process has been designed to convince the residents of this area that this process will endanger their health, safety, and possibly their lives. The hysteria being generated is clouding the issue."

Talk about clouding the issue: On the same day the advertisement appeared, so did a large, accidental cloud of smoke from Stauffer. A controversy ensued when a fire captain and Ron Novak, the air quality director in Hammond, had to wait forty minutes before anyone of authority for Stauffer finally arrived to explain things.

Novak said he found the plant gates wide open and the security guard asleep. It was not the type of incident that would bolster public confidence.

"There are so many people who think we're hysterical housewives —frustrated with nothing else to do," said Marge. "But there's *a lot* I'd rather do and that I *have* to do but don't get done because I'm transcribing tapes or at the copy center or at the typewriter letter-writing."

Lydia told me her involvement came from simple concern for her family. "I have to be able to live with myself. I have to know that I did everything possible to stop them. If some fatal health problem developed in my children, I have to be able to live with my conscience. I have to know that I tried my best to protect them."

The women eventually ran their own full-page ad, showing a mother holding a baby. "WE HOLD THE FUTURE," this one said at the top.

The mother, of course, was wearing a gas mask.

The women had banded together a year before, worried about some of the things they were seeing in the neighborhood. They made no claim to epidemiology, but nonetheless the lung cancer rate seemed high here, too, and so did leukemia and cancer of the brain.

By their reckoning, there had been nearly a dozen recent leukemias in the immediate three-by-eight-block community, and twenty-six brain cancers during the past twenty-five years: one in a house that was next door to two other cases, they said.

"You know who else died of brain cancer was Sue over there," interjected Marge, pointing just south of her own house.

This was all a bit intriguing, but mere word-of-mouth. Brain cancer is not a common cancer. The annual U.S. rate is under five cases per 100,000 people. There are only twelve hundred or so people in this neighborhood.

That meant, perhaps, that a number of these "brain" cancers were indeed cases in which cancers had spread from some other organ and were misclassified or otherwise wrongly reported by the untrained citizens. According to the state board of health, twenty-one residents of the county that includes East Chicago, Hammond, and Gary died of brain cancer in 1985.

Since the county's population is 498,000, the rate is actually somewhat lower than what normally would be expected.

In the atlases for 1970–1979 there are no extraordinarily high, countywide leukemia rates, either.

But the total incidence of all types of cancer was significantly higher than the national rate. So was the specific category of respiratory cancer, which, at 68.5 per 100,000 for white males, was a full notch above the slightly high occurrence in Midland, Michigan.

Colleen said puppies had been born with water on the brain, and I heard stories (reminiscent of the Chicago projects) about hair loss— including a little bald girl in the neighborhood. When residents complained that the air burned their faces or just plain stank, officials told them it was just rotting vegetation.

Marge, whose father had died of lung cancer and her father-in-law of a tumor between his heart and lungs, added that neighborhood dogs seemed to sicken after rolling in the grass. They broke out in skin irritations, vomited after licking their paws, and lost their fur.

Some days the air smelled like burnt firecrackers, and when she washed windows she noticed they were etched and grainy as if something had eaten at the glass.

Lydia, a rather formal yet quite vocal blond woman with twinkling green eyes, spoke of a stepfather who had died of leukemia. She also said three out of the four people who'd lived in the house before she moved in had died rather young—one of pneumonia, one of a heart that "exploded," and one of bone cancer.

Lydia had had a baby that was no sooner home from the hospital than he had to be hooked to a heart monitor, he was so premature. The child had since developed asthma and ear disorders.

The baby's birth was right around the time of the catastrophe in Bhopal, and that made Lydia wonder about the air here in northern Indiana.

She walked around the neighborhood and counted forty-two yards in which hedges looked sick. "I thought, 'Oh, my God, it's here!' "

Theirs is a modestly prosperous neighborhood in a run-down town. The homes are ranch or Cape Cod in style. Not too far away are streets that consist of union locals, pizza ovens, and broken, sparkling glass. Just over the border in Calumet City is an old strip of shabby go-go joints and corner taverns that once had hosted the famous speakeasies of Al Capone. Had the gangster still lived, he would have found himself replaced as public enemy number one by Stauffer Chemical.

Or at least that was the case in the minds of Lydia, Colleen, and Marge. Actually, Stauffer was not an especially notorious transgressor of environmental law, and when questions are posed to the firm, the folks at headquarters obligingly reply.

However, the company *has* had a few noteworthy problems. A fire erupted at one of its plants in Chicago Heights eight years before, raising alarm because it involved pesticides; and just that last April the plant in Hammond faced $69,605 in proposed fines for alleged violations of hazardous-waste regulations.

Back in my old haunt of Niagara Falls, Stauffer had disposed sulfur wastes at a section of Love Canal that had escaped national attention because it was at the end opposite where people were evacuated.

In Indiana, Stauffer was recovering used sulfuric acid from the refineries and recycling it. The company had similar plants in Texas, Louisiana, and California.

Sulfuric acid is just an ordinary criteria-type pollutant, but what made this process interesting was that for fuel, Stauffer had used waste solvents and other high-BTU materials contaminated with heavy metals and probably "hundreds of other substances in trace amounts," according to Novak of Hammond's air quality division.

In essence, waste materials were fueling its process.

Some called it "nonconventional," "alternate," or "supplemental" energy.

It was a creative burst of euphemism that steered far clear of the alarming term "toxic wastes."

At one time, Stauffer also had tried using coke tar.

What was coming out of its stacks was anybody's guess. The company had sought and had been granted a request that kept its emissions and control efficiency a confidential matter. And there was no way Indiana or any other state could currently monitor the air for all the compounds that might be floating from the area's industrial processes.

Inspectors could climb the stacks and take periodic samples, but there were limitations to this as well. "A stack test only tells you what happened during a stack test," noted Novak. "The whole damn thing can go wacky a minute later."

Now, with the market for its sulfuric acid going soft, Stauffer wanted to take its whole system a dramatic step further. To make up for lagging sales, it planned to turn its sulfuric acid furnace into a commercial hazardous-waste incinerator.

We were no longer talking about just "supplemental fuels." We were talking about large, concentrated volumes of highly toxic residues that Stauffer proposed to take from other firms and burn for a fee. On its application was a list of chemicals that stretched on for 354 lines. Among them was a Hall of Fame of dioxin and furan progenitors, including 2,4,5-trichlorophenol, 2,4-D, pentachlorophenol, hexachlorophene, dichlorophenol, and chlorinated benzenes.

About the only thing it didn't list was PCBs, the burning of which Professor Hites told us is the single most efficient way of creating furans.

"Second best, but almost as good, is if you've got a phenol—an aromatic [benzoid] with some chlorines on it and a hydroxy group," said Dr. Hites. "Then you can form either a dioxin or a furan by linking these two rings together. That's the first step. You've got the precursors already there, more or less. So it's a pretty easy reaction in a hot system, because you've got adequate heat and oxygen around."

Colleen, Marge, and Lydia were no experts on the combustion of aromatics, but they knew they didn't want a hazardous-waste incinerator several blocks away from their homes. They also knew that just two years before, a tanker of "supplemental fuel" had exploded while feeding the furnace.

Though it did not involve Stauffer, an explosion on September 4, 1986, had forced nearly a thousand residents from their homes in Elkhart—another poignant omen.

In a meeting to set forth their worries to Stauffer's plant manager, the three women had listened as Mr. Miles opened the session with a joke. This was intended to both break the ice and make a point.

It did neither. It was the story of a man who went to a psychiatrist because he thought he was dead. No one could convince this man that he couldn't possibly be dead. No matter what kind of proof they gave him that he was alive, the man just wouldn't believe them.

When the psychiatrist could think of nothing else to do, he pricked the man's skin and drew some blood—proof, if ever there was proof, that the man was a living being.

The punch line: Instead of being convinced he must be alive, the man concluded instead that "dead people bleed."

The analogy was obvious: It was like citizens who, no matter what they're told, refuse to believe that a process like Stauffer's proposed incineration is really safe.

Realizing that they were being compared to a psychiatric patient,

80

the women had failed to react to Miles's punch line—staring stonily at him.

"You're supposed to laugh," Miles said after a brief, embarrassing silence.

The plant manager went on to compare the risk of benzene to the risk of drinking soda sweetened with saccharin.

"Come on!" Marge wrote in a subsequent letter to the editor. "If I choose not to drink saccharin, can I likewise choose not to breathe benzene, or for that matter, chlordane, or formaldehyde?"

When the women mentioned a study which showed that maximum chemical readings would be within fifteen hundred feet of the stack—which would include a number of surrounding homes—Miles said the fifteen-hundred-foot figure had been an error and that when they were checked again the projections indicated there would be no significant contamination ten feet from the stack.

"We know that if there's an incident with a trailer, or material coming to our plant, we're going to be involved in it," Miles pointed out to them. "It's going to be Stauffer's name that's going to be on the front page of the paper."

"But," interrupted Colleen, *"we're* going to be dead."

In a stunning victory for the women, Stauffer announced on October 13, 1986, that it was withdrawing its permit application because it had become such an "emotionalized" issue. The company was also upset because the environmental group Greenpeace had climbed its 280-foot stack to unfurl a banner which said "POISON." That had caused a big community impact.

Colleen's involvement went back to her beauty shop, where she heard too many accounts of cancer from her customers. "They would talk about their husbands or their sisters or their neighbors; all of a sudden, it just seemed like everyone was coming down with it," she said.

In her middle forties, Colleen was a big, pleasant, and yet defiant woman, a tireless leader, tough and earthy enough to have once grabbed a would-be teenage mugger by the neck, tossing him away from her in a parking lot.

As far as Stauffer went, she and Marge had shown their mettle by selling candy door-to-door in order to pay for the mailing of twelve thousand protest letters, sent to various state and federal officials.

Colleen's beauty salon was in the basement of her house. There the women and other protesters congregated to talk about the steel mill odors and the fading, grit-coated cars.

It was also a base from which they set out to follow suspicious trucks. I watched several of them brave through a cloud that may have contained asbestos just so they could snap pictures of a load being dumped.

But mostly they talked about the politicians and what the women saw as a circus of double-dippers and general hanky-panky. The East Chicago City Council was a special obsession with the women.

Most flagrant among its zany ways was the fact that in addition to their council stipends, most councilmen had some sort of full-time public job. They could also pick up extra pay by serving on various community boards. The situation was such that although East Chicago barely qualified as a bona fide city, its mayor was expected to collect $73,100 in salaries during 1986—or more than the 1984 salaries for thirty-two of the nation's governors.

The city was apparently an unprecedented hotbed of sports activity, too, for altogether the school system had not one but *four* athletic directors—more than it had high schools. One was the brother of an imprisoned former county clerk. Another was Frank Kollintzas, a city councilman.

"This is America?" a citizen had once complained.

"This is East Chicago," a councilman explained.

When I went to observe a meeting of the council, which had been known to shut off the lights on protesters like Marge, Lydia, and Colleen (or, as it had also done, to take away a dissenting councilman's microphone), another councilman, Gus Kouros, quite literally rushed out of the room when I asked about his extra earnings as the school system's "food consultant."

Food consultant!

The door he closed in my face bore a sticker promoting an event sponsored by the American Cancer Society.

What upset the women was that these same councilmen were now trying to woo into town a Canadian company that commerically burned hospital wastes.

As if Stauffer wasn't enough, the city was recruiting a hazardous-waste incinerator!

In fact, so anxious were they to have this incinerator that the city's director of air quality issued a permit to the firm without holding a hearing for public comment.

That violated the law, claimed one state official, but East Chicago shrugged off such a charge. It once had been reprimanded by the state for illegally issuing a clean air permit to Inland Steel, accepting, "without evidence," Inland's assurances that the proposed expansion would not violate air quality levels.

The city's own trash incinerator was also a concern to the state, for it was already burning hospital garbage.

With all the plastic from syringes, bottles, and other disposable ware, which offers the requisite carbon and chlorine, the concern with the medical trash was once more the formation of dioxins and furans.

To make matters even less palatable, the firm wanting to build the new medical incinerator advertised itself as handling blood and tissue wastes, contaminated animal carcasses, urine and stool samples, chemotherapy and biotechnology leftovers—plus wastes from AIDS and hepatitis victims.

I gave a call to John Armenta, the air quality control director who had granted the permit. When I asked about his background, he told me he had worked as a "roller grinder" at U.S. Steel for thirty-seven years.

Armenta conceded there was still a problem with particulates in East Chicago but that, to his knowledge, there was no monitoring being done for substances like benzene.

Benzene, as I've mentioned, can cause leukemia.

How did he feel about Colleen, Lydia, and Marge's claims of just such health problems in town?

"I wouldn't agree with that. I don't see any higher rate of cancer here. I would say the air here is 50 percent better than it was fifteen years ago."

Had he reviewed the city's cancer statistics?

Well, no, he said, but he asked me to wait a moment while he consulted someone out of earshot of the phone.

When he got back he announced the city's cancer rate was no higher than other cities'.

When I asked who had given him that information, he said the person he'd just consulted was somebody who just happened to be in his office—a public relations man for Inland Steel!

While Stauffer and the hospital wastes were the focus of current community worries, there were literally dozens of other potential pollution sources. The corridor of Gary-Hammond—East Chicago is one of the nation's top ten stretches of industry, to my eye.

I was chauffeured around one day by an environmental activist named Blythe Cozza, who runs a management consulting firm and works for the Chicago Association for Commerce and Industry.

She is also an ardent Republican, a trustee for the Republican Presidential Task Force. Besides underscoring the fact that environmental problems are not, as people often assume, a liberal issue ("I'm to the right of Barry Goldwater," she boasted), Miss Cozza also hints at a new trend: the involvement of business types in fighting pollution. ("For real economic development," Blythe said, "you need a healthy environment.")

We passed truck terminals and a Union Carbide facility and an intersection that made me take a second look because it was called Hemlock Street. I had also noticed a New Jersey Avenue some hours before, a hint of things to come.

She took me into Calumet Township to see a man named Richard Murzyn who regularly protested the fumes coming from a nearby asphalt plant by blowing off an air-raid siren.

Murzyn, come to learn, was also a new type of environmental activist, and his credentials should have given the air polluters great pause. For Murzyn's nickname was "Mo Mo," and he had just finished five years in the federal penitentiary. In the newspapers he had been linked to a reputed hit man.

Blythe the right-winger returned me to the home of Colleen Aguirre. On the way I noticed a special bus with handicapped children making quite a few stops on one particular block.

After the council meeting, and after calming down Marjorie Surufka, who was so upset by the councilmen that she was weeping at the kitchen table ("I just hate them so much!"), Colleen settled herself and calmly, objectively began to explain how another possible victim of airborne contamination was her husband, Tony.

Just a short while after her interest in cancer had been sparked by what she was hearing down in the beauty shop, she learned that Tony was also now a statistic, afflicted with chronic leukemia.

84

"He was losing weight and he had leg cramps, and fatigue," she told me in a hushed tone (for Tony was trying to sleep down the hall). "He usually came home from work and cut the grass or washed the car, but he wasn't doing anything—was just really depressed. He looked at me and said, 'I don't think I'm going to live much longer.'

"And I knew there was something wrong because he wasn't the type to complain. This went on for two months. The first diagnosis had been that he had sugar a little high and his uric acid was too high."

She took him to another clinic for tests, and the results came quickly. His count of white blood cells was 490,000, where 5,000 to 10,000 is normal. There was hemorrhaging behind his eyes.

"We went to the lab and the x-rays and then to lunch and came home. When we got home my older son was here and he said, 'Mom, the doctor called and says to call him right away.'

"So I called back and that's when I found out. He said, 'Come back and bring your husband; he's very ill.'

"And I said, 'What's wrong?'

"And he said, 'Well, I don't want to tell you over the phone.'

"And I said, *'What is wrong?'*

"And he said, 'It's a very serious blood disease. Can you take it?'

"I said, 'Yeah.'

"He said, 'Tony has leukemia.'

"I was totally stunned. I actually couldn't believe it. All leukemia meant to me was death."

Her eyes suddenly moist, Colleen Aguirre was no longer the tough, unflappable environmental leader who shouted down politicians and went nose to nose with Stauffer Chemical. She was a woman whose voice had turned gentle and whispering and scared.

It had been awfully hard hearing the doctors say that the odds were Tony was going to die fairly soon and that because of the harsh medical treatment the disease would warrant, he might die very miserably.

After the diagnosis, she had called her best friend, Julie, "and then I cried really hard. I lost it for about three minutes there."

She'd known Tony since she was a girl of fourteen, met him at a carnival, and he was a high-school football player, a proud, healthy Latino who never needed a doctor, who wasn't even born in a hospital.

They'd had three children together, the youngest, Bobby, ten. She showed me family photographs of Tony sitting next to a Christmas tree and opening presents, still looking quite robust and not at all like the pained and wasting and wraithlike man who now roamed the household.

Having heard what the doctor had to tell him, Tony had held in his own tears until he could wander off alone to their car that dreadful day.

Later, while he was in St. Luke's Medical Center in Chicago, Colleen had trouble sleeping, so sometimes, at two or three in the morning, she would get into the car and drive the half hour into town to watch Tony sleep.

"I'd just stand at the door to make sure he was okay," she said softly.

While we talked, the youngest son wandered into the kitchen to help prepare his father's nightly injections of interferon, the new anti-tumor drug.

Bobby had been a sickly child, his lungs filled with fluid upon birth. ("For three days, " said Colleen, "we didn't know whether he would make it.")

But the boy was now a precocious fifth grader with ambitions to be a doctor. When he found out about Tony's leukemia he cried one whole day, but then he went to the library and got some books on the subject and learned to pronounce multisyllabic medical terms so he could be part of his father's fight.

He also had been brave enough to accompany his dad down to Houston and a cancer clinic there.

"I told him, 'Bobby, you're going to see some very horrible things,' " Colleen recalled. "There are people with growths on their face that are so big they lay on their shoulders. And amputees.

"I'd like to take these politicians and make them sit all day on the ramp at that clinic—it's like a chamber of horrors! People look like they would at a concentration camp. There was this one cute little woman with her face marked up by cobalt treatments. Later, the whole bone structure of her face was gone on one side.

"In the cafeteria, there were four or five boys who were bald and on IVs."

Colleen added that although the interferon had greatly weakened him, Tony still went to work at Inland Steel "because I think he figures as long as he can work he'll make it."

With a little help from his mother, who tapped some of the bubbles from the tubes, Bobby meticulously set about rubbing alcohol on a bottle, pulling sterile water into one of the syringes, and mixing it with different types of interferon crystals—gamma and alpha, two separate shots that take the boy nearly half an hour to prepare. Not wanting to cause a single furan, the plastic litter is kept aside by Colleen instead of being hauled off by the garbagemen.

The bottles were capped and the table wiped clean.

His measurements completed, Bobby—looking much too solemn for a ten-year-old boy—marched off to inject his father with the interferon.

Outside, the throttle of a locomotive split the vacant night. No doubt it was moving coal and ore, as well as a few tank cars.

The sounds of industry had a beckoning effect: It was time, before heading down the Superwhip, to touch on some other areas at the periphery of the Great Lakes—the insidious new fumes of old Smokestack America.

9

There is Michigan, of course, and besides the puzzlements of Midland, there is the expected big-city grit in Detroit.

Its waterfront, rising anew from the ashes of Rust Belt decay, still has a tool-and-die building, and an old foundry, and gray wisps of smoke from a Chrysler assembly plant.

According to the state, it would not be surprising to find eight million pounds a year of xylene, toluene, methyl ethyl ketone, and other solvents coming from an assembly plant. The coke ovens here are driving off substantial quantities of benzo-(a)-pyrene, too. At a GM plant upstate, PCBs had been a problem.

Meantime the city had broken ground for a waste-to-energy incinerator that could burn 6.6 million pounds of trash a day and will cost $500 million to build—the biggest such project, at the time, in America. Overall there were 88 such incinerators operating or being built, 39 about to start construction, and at the very least 106 proposals for other ones around the time of the groundbreaking.

Burning refuse to generate electricity has become one of the hottest new industries in America.

It is also one of the most controversial. Critics charge that dioxin will be released from the incinerator, which is in the heart of Detroit's major ghetto. The compound has been detected in similar burners located in Toronto, Ontario, and Hemsptead, New York. In Switzerland it was found that cows grazing near a municipal incinerator had dioxin in their milk.

Projections show that on some days the plume of the Detroit incinerator will move from the slums over the downtown to Grosse Pointe—the most affluent part of Detroit, crossing the estate of Edsel and Eleanor Ford. One report by the state suggested that Detroit, Windsor, and nineteen suburbs will face higher cancer risks.

From there it will move over Lakes Erie and Huron.

Let's say that the increased cancer potential would amount to twenty cases per million people, which is within the range of the more frightening estimates. That would mean there will be fourteen times the chance of getting cancer from the Detroit incinerator than there was of getting killed abroad by a terrorist during 1985 (an especially bad year for terrorism); and about twice that of dying in a hurricane, earthquake, tornado, or flood.

Living just downwind of an incinerator is safer than smoking a pack of cigarettes every day (that is a risk of four in a *hundred*), or for that matter than driving an automobile.

But it is more likely to result in death, on the other hand, than flying in a commercial airliner.

I had watched the groundbreaking ceremony as a number of rather rag-tag demonstrators stood in the rain thumping a drum and holding signs like, "We Won't Get Burned." One man told me he had been threatened and bodily removed by city henchmen from a public hearing.

Another time, another place (far to the west in Montague, on the Lake Michigan side), I had spoken with a sixty-nine-year-old man who sometimes had to drive away from his home and sleep in his car because of the fumes that had come from a Hooker Chemical Company plant. An affidavit from a former employee claimed the factory once had vented chlorine and C-56 (a chlorinated pesticidal ingredient with threshold exposure limits ten times more stringent than phosgene nerve gas) into the neighborhood air. He said he had been instructed to "act as though the white vapor escaping from said pipe was steam."

In St. Clair County, which is across the river from Sarnia, Ontario —Canada's "Chemical Valley"—the cancer rate had jumped more than 20 percent between 1980 and 1984.

Sarnia is like a transplanted chunk of East Chicago. And its effect on the lakes is surely comparable to any cotton-field pesticide. There are fourteen major petrochemical plants in the area, and the river coursing by them, the St. Clair, is loaded with toxic compounds, among them perchlorethylene, which had formed blobs at the bottom.

In fact the pollution is so bad it had caused the face mask of a scuba diver to actually disintegrate.

Dow owns the plant being blamed for the caustic blobs. This factory leaked thirteen thousand pounds of chemicals, including seventy-seven hundred pounds of vinyl chloride, the very day after I passed by its front gate—more than four times the amount of vinylidene chloride Dow had said it released in an entire year down at its Freeport, Texas, plant, where there had been allegations of high brain-cancer rates among its workers.

Understandably, there is now a movement afoot to change the nickname of Sarnia from "Chemical Valley" to "Happy Valley." And to change the country's ten-dollar bill. For as long as anyone cared to remember, the currency has proudly displayed a panoramic picture of Sarnia's refineries—the largest cluster of such industry in all of Canada.

But now, with the blobs, with the site of smokestacks less welcome than in former years, they were talking about redesigning the money with the picture of a simple bird.

(Would the bird have good feet and a straight beak? This is Happy Valley: of course it would.)

Sarnia is but the most obvious manifestation of toxic releases that occur all across the Ontario province. From Chemical Valley through Hamilton, Mississauga, Toronto, and the rest of the lower peninsula, to jump quite a bit east for a short while, is an assortment of industry very similar to that of Ohio, Pennsylvania, and New York, which are on the other side of Lakes Erie and Ontario.

On a lonely island in the St. Lawrence River, straddling the boundaries of Ontario, Quebec, and New York, Mohawk farmers have reported symptoms among their livestock quite similar to fluoride releases we'll see in Montana. Here it is a Reynolds aluminum plant that gets the blame: Over the past two decades 25 million pounds of fluoride have fallen on the island, and once-majestic pines have turned into huge toothpicks.

"Overall, about 50 percent of the acid rain affecting Canada originates in the United States," says a document put out by Environment Canada, the nation's version of EPA. "About 43 percent of the two million lakes in Ontario and Quebec, as well as 40 percent of the productive forest area in Canada, are located in areas that receive moderate to high levels of acid deposition *and* are at least moderately susceptible to damage."

Canada was so desperate for the ear of President Reagan that it had hired (it was charged) his closest former aide, Michael Deaver, to lobby for emission reductions.

It can be argued that Canada and the rest of the Northeast were overstating the role of the Midwest, Ohio Valley, and Tennessee Valley in causing acidic fallout. When I paid a visit to the Geophysical Fluid Dynamics Laboratory in Princeton, where one of the most powerful computer systems in the world, known as "CY4," works overtime, Hiram Levy II, a chemist there, showed me new evidence that the role of actual *precipitation* in acidic fallout is being significantly exaggerated and that much of the acid nitrogen that falls to the ground does so in *dry* weather and comes from *local* and *regional* sources which haven't been taken into fair account—not just far-off smelters.

For the province of Ontario, however, states such as Ohio and Pennsylvania *are* in fact a "regional source." Intense in chemical manufacture, they without a doubt contribute not just sulfur from the tall power stacks, not just nitrogen from their cars, but also the barely mentioned vapors and fine particulates carrying more toxic compounds. Deposits of cadmium, once thought of as a product of the earth, are now also a product of the sky.

Into every life a little rain must fall, but in Mt. Forest, in Windsor, in Burlington—did it have to carry traces of chlorinated hydrocarbons with it?

PCB-like material was being tracked in human fatty tissue and mother's milk from Kingston to Ottawa.

Down in the easternmost fringe of the Midwest, Ohio is best known for iron and steel, and for the acidy, coal-fired furnaces. According to an estimate floated in journals such as the *Bulletin of the Atomic Scientists,* each such power plant, spewing trace metals, spewing polycyclic aromatics, spewing oxides that swell the tissues of the lung, can cause fifty deaths or so a year—more of a worry, in some ways, than a nuclear reactor.

But, that should not overshadow the state's simultaneous and equally important involvement directly with chemicals themselves. Ohio is very big in rubber and plastics, two fields that should always generate special concern. In Akron, a Goodyear plant was releasing 63,619 pounds a year of acrylonitrile, and Standard Oil in Lima, 95,700 pounds. Used in the manufacture of synthetic fibers, acrylonitrile is every bit as toxic in the air as benzene, and probably significantly more so.

Throughout Ohio, gases rise from stacks, from vents, from sewers and transport points, from scrubbers, from product dryers, from storage tanks and holding ponds.

In Vickery, a small town near Sandusky halfway between Toledo and Cleveland, there is the Waste Management site where huge ponds of chemical slurries had filled the air with a caustic mist. "They had five or six lagoons, each about half a football field in size," says Al Franks, spokesman for the Ohio Environmental Protection Agency. "Just driving by there was horrible."

In the shadowy outskirts of Toledo whatever is in the air seemed to be peeling the paint of buildings there. Sulfur was lying all over the ground, and rotted barrels caused fugitive fumes. "FOG—SLOW AREA," said a sign near a refinery.

If we were a fine particulate traveling at thirty-one thousand feet (behind a jet plane spewing more oxides), we could look down on Ohio and discern the checkered patches that are farmland, the crosswork of roadways, the snaking creeks. And also the threads of smoke from Columbus—one of the least industrialized of major Ohio cities—that serve as a mild prelude to the heavy-duty steel, chemical, and automotive plants of Cleveland and also Cincinnati (where a plant owned by Morton Thiokol was said to be releasing 400,000 pounds of organic chemicals each year).

How about all the formaldehyde from other plants scattered about the state, or the organic vapors from a phenol process at a U.S.S. Chemicals factory in another iron town named Ironton?

With just the right timing, and with chromatographic eyes, we could have seen the lead floating around Delta, Ohio, or residues from a uranium processing plant (south of Columbus) that has leaked hexafluoride. People living near yet *another* uranium plant (this one north of *Cincinnati*) already have shown radon in their bodies.

Or we may have glimpsed the white phosphorous that caused more than twenty-five thousand people near Miamisburg to evacuate when it escaped from a derailed car.

But no litany of horror stories is necessary to realize the very significant contributions from the Buckeye State. One has only to know that in addition to its manufacture of metals and machinery, Ohio is the nation's fifth leading state in the value of its chemical shipments—a spot ahead of California and just below Illinois.

Did Canada, did the Great Lakes, did America's East Coast not expect something to arrive from Ohio when, based on 1982 figures, it did about $9 billion in chemical business a year?

If there was any lingering doubt about long-range transport one had only to order up from the National Oceanic and Atmospheric Administration a report entitled "Cross-Appalachian Tracer Experiment." The experiment consisted of releasing the tracer compound perfluorocarbon in Dayton, Ohio, and Sudbury, Ontario, and tracking it by ground samplers and airplanes.

The trace compound was easily detected at points up to 687.5 miles away—from Virginia up through New England and southeastern Canada.

PART II

DUST IN THE WIND

10

One way to get a furan or dioxin way up to the fast winds is a large, convective thunderstorm.

Dust and all manner of other particulates are not just stirred around but actually sucked up in such a weather system.

If the spiral of upwind becomes violent enough, and touches the earth, it is known as a tornado.

It was the spring rainy season, and I was up in northern Minnesota to scout the region where the top of the Gulf Superwhip slams into the Canadian Cool.

To lend yet more karma to such a convergence of natural forces, this is also where the Mississippi River begins.

The river's source, a five-hour drive from Minneapolis, is in the mysterious area where Paul Bunyan was said to have felled his hauls of timber, and where the "abominable snowman" is also part of the local lore.

At roadside were mirrorlike lakes filled to the brim, and fawns scampering for cover.

As I drove toward the headwaters, flank and wall clouds obscured a puffy, towering cumulonimbus thunderstorm machine which now claimed the horizon. Its updrafts and downdrafts were probably approaching twenty-five miles per hour.

Lightning darned a suddenly ominous sky, which was now a very deep blend of gray and blue.

As I fought through the downpour, tornado warnings were urgently announced on the radio.

The rain was becoming too intense to drive. Cautiously I made my way through an Indian reservation and wildlife refuge. A mile away, said the announcer, residents had spotted a funnel cloud.

Perhaps sixty thousand feet above, the rotating updrafts, if they were now tornado strength, were overshooting the top of anvil-shaped clouds. That implied that they were injecting surface air into the stratosphere.

Tornadoes are so powerful they have been known to drive a straw through a metal pipe or lift a rail car. I glanced into the rearview mirror, afraid I would see a funnel sneaking up the lonely back road.

I looked for cover. There hadn't been another car for what seemed like hours.

Finally I came to a general store. It was the Lake Itasca Trading Post, and I hurried inside.

There I found Al Katzenmeyer, an old beaver trapper wearing a cap and overalls, his jowls bristly, his eyes slightly crossed. As he chewed tobacco and checked his mail he explained that a tornado or two had passed very near the general store in recent years—one right out front, added the woman at the counter.

When I asked how they knew when one was coming, they told me the sky would turn a peculiar shade of green, and thunder would volley.

The rain eased and Al took me into a hilly forest of balsams and aspens. The land was plentiful with partridges, woodpeckers, grackles, and meadowlarks. Nearby was a bald eagles' nest.

He led me down a path out of the thicket and there it was, a brook gurgling as it eddied over several dozen field rocks at the outlet of Lake Itasca's northern arm.

It was knee-deep, the Mississippi, and only fifteen steps across. I crouched to stick my hand in the water at this very point, 1,477 feet above sea level, where a river that eventually drains 41 percent of the mainland United States—carrying more than 100 trillion gallons of water a year to the sea—so modestly begins.

By the time it ends it will have swept past the most treacherous turf of Toxic America—the base of the Superwhip.

On the other side of the rocks two loons floated on the lake and wild rice poked up between them. The water was cool and clear, and the rice attested to its purity, for the reeds cannot germinate in waters with a sulfate ion concentration greater than ten parts per million. It was not for nothing that they called this Clearwater County.

But I was to learn that other lakes in the northern part of the state, however clean they may appear, have encountered a perplexing problem with one of the oldest of commonly known pollutants—the toxic metal, mercury.

Shiny and globule in liquid form, mercury exists naturally in the soil and has been used in a wide variety of processes, from pulp operations to farm fungicides. In Europe, the term "mad as a hatter" came into being a century ago and referred to the neurological and behavioral changes mercury can inflict upon production workers who come into unfortunate contact with it (in this case, by using it to soften the fur in making felt).

Lewis Carroll picked up on it, and thus gave us the Mad Hatter of Alice in Wonderland fame.

Though it takes forty thousand times as much mercury as it does dioxin to cause similar alarm in food, mercury is no toxicological slouch: kidney and intestinal damage can also follow in its wake. When it is found in fish at more than one million parts per trillion (a single part per million), it's considered a significant problem for the humans eating them.

One of the most famous calamities associated with mercury occurred in Japan during the 1950s, at Minamata Bay. There, near a plastics plant, birds fell from the sky and sickened; cats roamed the streets, whirling and leaping spastically.

It is what a clinician might call behavioral abnormalities.

Whether ingested through contaminated fish or inhaled as a vapor that diffuses readily through the lung's alveolar membranes (which bear certain resemblances to a sponge), mercury binds to red blood cells and eventually finds its way to organs such as the brain, causing cells to shrink and then convulsions that can lead to a torturously dizzy death.

At Minamata, fishermen and their families suffered blurred vision, numbness, and progressively worsening symptoms, including grotesque limb contortions and birth deformities.

There were no dancing cats in Minnesota yet, but the metallic contaminant already had been linked to poor courting behavior and sickly reproduction in loons eating from the Wabigon River. Methyl mercury, an especially troublesome type of the metal, was collecting in the fish at levels only four to eight times less than that found in the poisonous fish at Minamata.

99

No one knew where it was coming from. One theory was that acid rain in the region chemically liberated the mercury from natural soil deposits, and the mercury—turned into a more mobile organic form by bacteria—moved quickly into springs, runoff, and ultimately the beleaguered lakes.

But there was also reason to believe the major source of mercury might be directly from the air—as a sort of a metallic fallout.

Mercury has long been known to glide from chemical factories, copper smelters, and coal-burning power plants. There are smelters north of the border, in Canada, and there had also been a chloralkali plant up there which had severely contaminated an area near Saskatoon.

Minnesota, for its part, has a number of power plants.

From the south, where the Gulf air whips up hundreds of miles before moving east, there were any number of suspect sources, among them Texas and Missouri. It was known that they contributed acid rain to Minnesota, and if acid, why not some mercury as well?

As the beaver trapper and I left the Mississippi's headwaters, new thunder echoed in volleys through the dense wilderness. And flank and mammatus clouds once more roiled in the sky.

Ironically, several dark pouches of mist moved in a formation that bore some resemblance to the sockets and nose cavity of a skull.

There was also the greenish hue which I had been told forewarned of a tornado.

High above, apparently, the hail was strangely reflecting and refracting the light.

Big, cold drops of rain were falling, and though it wasn't twisting yet, and never would show me a tornado, once more the wind was sweeping to the east and forcefully drafting down.

In drier weather the main source of toxic matter, in many areas, is probably the pesticide-soaked dust that blows off the farms.

This is true throughout the central states.

In Minnesota alone 180 billion pounds of soil is lost each year because of wind erosion.

The figure is a startling four trillion pounds (not counting federal land) nationwide. Most settles out quickly, but some of the particles are sucked up to the jet systems or ride a wave of low-altitude air.

In certain cases such a ride might go on for hundreds of miles. Back in Michigan one year, an analysis of dust characterized some of it as having come all the way from Oklahoma.

The question: Precisely what kind of toxic compounds are in the curtains of farm dust?

What is being sucked up during thunderstorms?

Here in Minnesota, as throughout the region, farmers are increasingly dependent on herbicides, which now account for about 60 percent of the pesticides sold. There are about one thousand active pesticidal ingredients registered for use in the Midwest and across America—applied in some forty thousand products. Since 1964 their use has fully doubled, to an immodest 108 billion pounds in 1984.

By the end of 1986, according to an economist at the Department of Agriculture, the sale of pesticides was expected to total nearly $4 billion annually.

The DDT-type insecticides of the past—chlorinated hydrocarbons that bioaccumulate, or build up, in the body's fat—largely have been replaced by compounds that more quickly break down both in the environment (through action of oxygen, water, sunlight, or microorganisms) and in the body (where they are unhinged by enzymes and carried to excretion).

Instead of lasting decades, as some chlorinateds can, today's popular brands—at least in theory—last a matter of weeks or months.

Though it is always risky to generalize when it comes to insecticides, those in widest use today are for the most part composed of molecules with a closer resemblance to nature's own amalgamations. They are therefore easier to break down.

There isn't nearly as much chlorine, for one thing. True to its name, the highly popular class of modern pesticides known as "organophosphates" has a phosphate as a central molecule. Phosphorous is naturally found in a cell's protoplasm.

While, in some cases, benzoid rings are involved, the oxygen atoms in organophosphate pesticides are often replaced not with chlorine but with sulfur, which switches place with oxygens in natural reactions all the time.

An important factor is a molecule's electrical charge. Likes attract likes. The chlorinated pesticides tend to have a nonpolar charge and so have their great affinity for body fat, which is also nonpolar.

The addition of chlorine must always draw our attention. It changes a molecule's electronic configuration. Polar can quickly be made nonpolar: In the case of the pesticide chlordane, which is widely used in

101

termite control, only 2.5 parts per million in food can quickly build to thirty times that quantity in our bodies.

The organophosphates, on the other hand, are more polar and therefore prefer water and similarly polar fluids. That means they dissolve somewhat more efficiently, especially when a body acid latches onto their side rings.

Anything that dissolves is more prone to find its way through the blood and out the bladder.

Even if it stays in the body, an organophosphate is no longer such an active poison once its phosphate bond has been split. The body has enzymes known as esterases that are up to this job.

Nature also has an easier time recognizing two other major classes of insecticides, the carbamates and pyrethroids. There are some chlorines and fluorides in synthetic pyrethroids, but the central molecule, an acid ester, is very much like a natural insecticide found in the chrysanthemum. A carbamate, as its name implies, is structured upon groups of carbon as well as nitrogens and oxygens.

The hitch is that anything that harms beetles, worms, or the many other agricultural pests tends also to have toxicological implications for man. All animals share common biological building blocks, and though quite different in size and complexity, systems of respiration, digestion, and nerve conduction are present in both worms and man.

"No one really knows what he's talking about," commented one candid EPA researcher when I asked what the chronic effects of these current pesticides might be. Though far less a carcinogen than chlordane and other chlorinated hydrocarbons, any pesticide is probably prone to affect that wondrous enzyme factory known as the liver either in acute doses or small quantities over a long enough period of time.

Most obvious, as far as organophosphates, is their effect on nerves. Destroying a protective enzyme known as cholinesterase, they can cause an imbalance in the chemical transmission between nerves, resulting in an uncontrolled firing of nerve impulses that manifests as trembling, spastic muscles, and outright convulsions.

Because it is rather readily excreted, it takes in the neighborhood of six times as much parathion (one of the most poisonous organophosphates) as mercury to raise official alarm when found in fish.

Though the calculations can vary widely (and even wildly) depending on what data are plugged in, for rats the acute lethal dose of parathion compared to dioxin would be the same as $8.30 versus four cents.

However, this means if you divide an ounce of one of the most poisonous organophosphates, parathion, into 250 tiny piles and ingest a single pile, that probably would be more than enough to paralyze you before you could reach for the phone.

If the dosage is quite low, the person exposed to such insecticides may notice nothing more than stomach cramps and blurry eyes.

As with most other synthetic compounds, they also cause the standard and even vaguer toxic symptoms of nausea and headache.

In 1985 it was reported in an obscure science journal that researchers from Texas A&M University had tested for the mutagenic potential of agricultural soil on bacteria and had found that all the extracts indeed showed this capability.

Bacteria is the common barometer to gauge a mutagen. And a mutagen is a substance which can cause abrupt changes—mutations—in the genes of an organism, to be passed on to the next generation.

One of the soil extracts was only one order of magnitude less mutagenic than the condensate from cigarette smoke.

Quite simply, mankind had turned soil—the very food of vegetal life—into a mutagenic muddle. Much of the blame must fall not just upon the spraying, dusting, and blowing of modern insecticides, but upon the application of herbicides.

Of the billions of units in the DNA of a person's genes, it only takes one or two, changed or critically set out of its proper place, to cause a threatening disease. Our bodies operate upon principles of biochemistry, and so chemicals intimately react with us.

Before we get back to Minnesota, the northern Superwhip, and the already mushrooming possibilities of what might be circulating above this "unstained" region—in this massive respiratory system that blows upon much of America—it's important to note that as farmers have sought to prevent soil erosion and save on fuel costs by killing weeds with newfangled chemicals (as opposed to dragging a till back and forth), they have veritably soaked much of the nation's ground with herbicidal mixtures that, like the old chlorinateds, are suspected of causing birth defects and cancer.

Prime among these is a corn and soybean herbicide called alachlor, which attacks broadleaf and annual grasses. Its basic backbone is a

103

benzene-type ring with a nitrogen group and dangling side chains. In 1982, the last year for which agricultural officials had a solid number, 84.6 million pounds of its active ingredient were used.

Less toxic but of wider use was another of the grassy weed killers, atrazine, spread on 47.9 million acres of not only corn, but also grain sorghum. In dogs fed 1,415 parts per million it has caused only decreased hemoglobin and lower body-weight, but there has been no final determination on its cancer-causing capabilities, and herbicides of this class (known as triazines, also with a ring similar to benzene but for the nitrogens replacing certain carbons) cause concern when concentrated in fish at about the same level as parathion.

It takes a lot more of atrazine to immediately kill a rat, but other triazines are more poisonous than parathion over the long term, rife with suspected carcinogens.

Remembering that synthetic fertilizers add an extra little nitric bite, and considering the vast quantities (enough methyl parathion was spread around one recent year to hypothetically kill the world's population eight times over), the question of what is in the blowing dust—the whirling topsoil that can nearly blind a rural passerby on a windy Minnesota day —is certainly not an idle one.

From preliminary indications, in fact, farm dust would appear to be a major, unrecognized source of toxic air pollution, dispersing weed killers and fungicides that can cause everything from relatively innocuous hay fever–like symptoms to fetal death. In one of the very few studies done along this line, researchers from the U.S. Department of Agriculture (DOA) found up to twenty thousand parts per trillion of herbicides and organophosphate insecticides in the fog hovering over two test areas: Beltsville, Maryland, and the Central Valley of California.

Since the same readings could be expected anywhere in the expansive agricultural zones of the central states (and anywhere else, for that matter), the DOA findings should be heard as a clangorous warning.

Dr. Dwight Glotfelty of the DOA informed me that the volume of farm chemicals evaporating directly into the atmosphere—not counting what escapes on windblown topsoil—can range from "insignificant" levels to more than *half* of what is applied.

It can happen during the mixing, storage, or after application of a pesticide. "This vaporization," he added, "is invisible. It depends on a chemical's structure and characteristics and how it's applied. I'm not sure anyone's doing monitoring now."

Among those found floating in the sample fog were triazines and another class of herbicides called chloroacetanilides, he said, along with some chlorinated hydrocarbons.

So here, where America's nourishment is sown, and where the skies often fill with thunderclouds, where the very Mississippi begins, agriculture seems to be competing with industry in shoveling compounds skyward and across the plains.

11

To leave the weighty topic of chemical-heavy dust was to pursue once more the ephemeral industrial vapors that have long been ignored by government regulators who apparently had assured themselves that such substances disappear to near-nothingness in the seemingly endless ocean of air.

In Minnesota, as in the majority of states, the first real inventory of potential toxic polluters was conducted only recently (in 1985), but the state is not all that confident that the initial data represent a complete panorama of dangerous fumes. About one thousand companies were sent a questionnaire that listed chemicals of particular concern—a sort of checklist—and only two thirds of these firms responded with the requested information.

The state—an environmentally progressive one—wants to know what is entering the atmosphere in addition to the rather well studied, traditional pollutants such as sulfur dioxide, particulates, nitrogen oxide, ozone, lead, and carbon monoxide, all of which are known as "criteria pollutants" because they—unlike more toxic substances!—are regulated under the Clean Air Act.

While the upper central states from Montana to Wisconsin are not overly industrialized, many sources throughout the region are capable of releasing significant quantities of unregulated carcinogens.

If a factory is releasing TCDD in Minnesota (and most other states), for instance, there would be no specific air code to stop that. Action would have to be taken in the permitting process or through the official (and very unlikely) declaration of an "imminent hazard."

Nor is there a comprehensive and regular monitoring program to

107

detect what other hazardous chemicals are in the air. Except for an incipient program in New York, no states have set in place a continuous, statewide network of toxic air monitors.

As far as ozone, which *is* monitored and can result from any number of free-floating hydrocarbons, Minnesota, for one, has trouble in meeting federal criteria, which consider a tenth of a part per million the maximum acceptable level.

None of this was much to the liking to a former garment salesman named Leslie Davis, of St. Paul, who ran an organization that sounded a bit like a Saturday morning cartoon hero—"Earth Protector, Incorporated"—with Davis as a sort of Clark Kent.

At forty-nine, Davis was highly unusual as far as environmental activists go. Most are small-city housewives who work from home, in the tradition of a Colleen Aquirre. Davis, a native New Yorker, had been a wealthy businessman who fought through the ranks of the Manhattan garment trade. He had moved to Minneapolis in 1962 and became interested in environmental causes after reading an article in a dentist's office on nuclear wastes.

Davis's seriousness in the cause could be measured, in some part, with dollar signs. The former salesman had spent twelve thousand dollars of his own money to stop local utilities from burning waste oil contaminated with PCBs.

He worried about dioxins and furans, and the bureaucrats were forced to listen. Davis was prone to hiring his own chemists and attorneys. Once, he had taken a matter before a United States appeals court. To make matters worse, he openly criticized the environmental officials "for acting like an arm of the industry."

Davis was also unorthodox in the way he had set up his organization. It was not incorporated as a nonprofit group, as most environmental groups are; clearly he planned to parlay it into some money. While his life-style had changed substantially in recent years (he was living in an apartment downhill from the affluent St. Paul neighborhood where he once owned a ten-room house with 2.5 acres around it and a rustic stone fireplace), he hoped to make a new living marketing sweat shirts, totebags, and stuffed animals based on environmental themes.

One of the stuffed animals was a beaver named Paddle-Tail.

There were also plans for coloring books with water-soluble paints (instead of harmful solvents) and animated cartoons: Someday he indeed wanted his stuff on Saturday television, between the cereal commercials.

It was difficult to keep up with all of Davis's causes, but suffice it to say they well represented the many types of ecological concern that can be found throughout the midwestern states.

Prominent among them were the four coal-fired electrical utilities in the area and plans in downtown Minneapolis for a plant that would burn garbage in order to generate steam-driven electricity.

He was also concerned about an incinerator that was owned by the 3M Corporation and burned hazardous wastes south of St. Paul. And with a coke refinery down there, too. I noticed in a federal document that there also had been some chlorophenol production in St. Paul, which raised the question of dioxin residues.

Of late, however, Leslie Davis was making headlines fighting with the Ford Motor Company. And he was giving Ford fits. It was the issue he had taken before the court, and it involved a decision by the EPA, Governor Rudy Perpich, and the state's Pollution Control Agency to grant Ford a waiver that allowed the company to emit more pollutants per vehicle than otherwise would have been tolerated.

By February 11, 1986, Ford was found to be exceeding even this lenient waiver. Davis—now out of the phone booth and in full regalia—charged that, as it was, the waiver would allow 400,000 pounds of volatile organic compounds to float over St. Paul and Minneapolis each year.

Already there were probably trees dying near the site, Davis fretted, and in the local newspaper had been an account of people, "who go about their daily chores in gas masks."

One sixty-five-year-old woman complained that she was going to have to move because of late-autumn fumes that invaded her apartment and then quickly fled, like a toxic cat burglar.

It smelled like a "burnt fuse," she explained, hinting that it was behind recent bouts with asthma, which had caused the widow to spend Christmas Day in a hospital.

There were no epidemics of three-legged chickens or, that I could find, any geese with their wings on backward. But the area of Canada and the United States that is summarized by the convergence of air in and around the upper central states is more interesting than one might initially suspect.

Minnesota and the state just east, Wisconsin, could point fingers at bordering territories. As it does in eastern Wisconsin, Chicago's blue haze can also find its way into Minnesota; and air coming from directly south—up the Mississippi—brings a kaleidoscope of pollution from the troubled states in the stem of the Gulf Superwhip.

These states are as important as any in the nation when it comes to hazardous air production, and we will look at them in more detail in subsequent chapters, for the portrait in the end is an alarming one.

But for the time being, it is well to remember that much of the flow in these central states also comes from the west and northwest.

To the north, of course, is Canada.

It's unfair to think of Canada's gargantuan, sparsely populated provinces as any kind of polluted mess. They include massive tracts of wilderness that are inaccessible to the automobile. The province serving as part of Minnesota's north border, Manitoba, has just over a million residents living in an area close in size to Texas.

But neither is it fair to believe that Canada is completely innocent of contributing to the synthetic chemistry of the continent's atmosphere. Besides the metal smelters that emit not just mercury, not just oxides, but also toxic lead, Manitoba and the two provinces to the west of it, Saskatchewan and Alberta, have paper mills, cement plants, ammonia manufacturing, chlorate units, amines, oil refining, pesticide formulators, steel, ethylene processes, mammoth fertilizer facilities, all manner of coal-fired furnaces, and potash.

Most of the industry is within two hundred miles of the American border. In one corridor near Edmonton, where Dow and B.F. Goodrich have facilities, vinyl chloride has been tracked in the air.

Canada's coal-fired furnaces, like the many found in the states just south, have, as anything burning coal or making coke has, the potential to release an extremely important contributor to any toxic stream: the potent set of polycyclic pollutants best represented by benzo-(a)-pyrene.

This form of pollution was at the root of the first investigation of carcinogenesis in man. The year was 1775 and a British surgeon, Percivall Pott, announced in a treatise that he had found excess cancers in the scrota of soot-covered chimney sweeps. Early in the current century, Japanese investigators discovered they could reproduce cancer in animals by rubbing their skin with such tar.

Another product of combustion, benzo-(a)-pyrene is also in the same category as furans when it comes to its pervasiveness. But it is far more readily detected than furans. Coming too from coal piles, wood burning, airport fumes, refuse incineration, and most especially tarring operations (not to mention cigarette smoke), it has all the earmarks of a classic carcinogen, producing tumors in all nine species it has been tested upon.

While not nearly as toxic as furans, in the air it is about 15 times as hazardous as regular benzene.

One must also remember the interactions and various additive effects known generally as "synergism." Though the Canadian sources dilute themselves rather quickly, traveling over raw wilderness, these little puffs, these stray molecules, are added to the emissions from Montana, North Dakota, and other seemingly carefree states. And together they add to the rivers of air like the tiny brooks that, in the end, form the great Mississippi.

The interaction of compounds—hundreds of toxic compounds, even in seemingly remote territories—is one of the modern age's trickiest and most pressing mysteries. For more often than not, we have no idea what the effect of two newly combined (or simultaneously inhaled) chemicals will have on our bodies, to say nothing of the collective effect that must be created when *dozens* of them intermingle.

In its own way it is a force with the magnitude of a relentless river. Benzo-(a)-pyrene is a fair example. According to one of the few pioneers in this field, the late Dr. George L. Waldbott, phenols, for one, can increase the carcinogenic power of benzo-(a)-pyrene. Another researcher has worriedly pointed out that on the skin of laboratory mice, the cancer-causing effects of low levels of benzo-(a)-pyrene are increased a thousandfold with the use of a solvent, dodecane, which is not thought to be carcinogenic by itself.

As long ago as 1928, experiments revealed that mice got skin cancer eight times as quickly when an exposure to tar was added to an exposure of ultraviolet light than when the light alone was used to cause the cancer. There are other indications that chemicals in the soil multiply the effects of benzo-(a)-pyrene.

When sulfur dioxide is reacted under the right conditions with benzo-(a)-pyrene, the result appears to be another example of an increased carcinogenic effect.

Indeed, any number of other organic compounds seem to become more toxic when exposed to the acidic conditions created not only by sulfates but also the nitrates.

And the westerly air heading into the top of the Superwhip is destined to carry plenty of these.

In Montana and the Dakotas, there are power plants, smelters, and productions involving sour gas. Add to this the forest industry in Montana, which burns logging slash and other wastes in teepee burners or in the open air, thus creating benzo-(a)-pyrene; the phenolic herbicide used on wheat fields in North Dakota (2,4-D); and the degreasing agents—the toxic solvents—that volatize in South Dakota from processes as diverse as making medical adhesive tape and heavy machinery, and you have quite a blend.

It is like a piggy bank: the air collecting beneath cliffs and rimrock, the wafts from little fires joining those from industrial furnaces, the pennies getting caught in inversions and quickly piling up.

Montana is a particularly good example. The environment there yields arsenic, nickel, zinc, sulfates, and cadmium (a soft, silvery metal that collects in the kidneys and can cause vomiting at under fifteen parts per million).

The prime concern, as always, is cancer and genetic repercussions. In the counties of Deer Lodge and Silver Bow, the death rate for lung cancer during one recent period was nearly *twice* the national average. As a state report pointed out, at least five of the metals found in Butte and Anaconda are considered to be carcinogens.

At the same time, eleven of eighty children tested in Butte were found to have high levels of mutagens in their urine. It was a finding that came as a jolt to worried parents who went to the dictionary only to find *mutagen* just above the word *mutation,* which meant a new character or feature that appears suddenly in plants or animals and can be inherited.

Under *that* was the word *mutant,* which conjured ugly images from science fiction.

In fact, in Missoula, Butte, and Billings (where there are sluggish winds and more rimrocks and more inversions), the mutagen levels closely resembled those found in some urban parts of Louisiana and New Jersey—the nation's prince and princess of toxic air pollution.

In Deer Lodge was a veterinarian named Dr. Paul Bissonette who described the effects of airborne fluoride. Every so often, he said, one plant near Butte—Stauffer Chemical—went around examining the cattle and compensating ranchers who claimed its pollution was damaging their herds.

Some of the ranchers, it seemed, were not beyond earning a little extra income by importing "junker cows" from out of state and presenting the aged animals as fluoride victims. ("The old gummers," said the amused veterinarian, "were twelve or fourteen years old!")

But Dr. Bissonette also said the problem was a real one. While fluoride strengthens teeth at levels below a part per million, it causes the exact opposite effect in excessive amounts. Teeth get soft and wear down rapidly.

Since cattle eat huge amounts of vegetation, which collects the wind's fallout, they are also prone to other symptoms of fluoride poisoning—gastric upset and calcification of the joints. Years before, said Dr. Bissonette, a smaller factory in Garrison had caused some frightening maladies.

"These cattle would go down, too sore to stand up, and they had to walk on their knees. The enamel wore off their teeth and the nerves were exposed and they had to lap up water like a dog would.

"One time I had a beaver skull and the teeth were worn down and *blunt.*

"Can you imagine the frustration of a beaver like that, trying to chew down a tree?"

The Mississippi is a big-time river by the time it reaches southern Wisconsin. On the way it widens into a slower-moving pool called Lake Pepin.

When I stood beside the lake it was brown with farm runoff—choppy and dark.

Darkness was just the right mood. More thunder came from the distance, and though this side of Wisconsin has little industry, I was to learn that the fish of this lake—as well as those in the Milwaukee River, Green Bay, and the Wisconsin River—are no longer fit to eat.

In their flesh, collecting like warehouse crates, are parts per trillion of a compound that society is going to have to confront urgently and without further delay—the wildly toxic, understudied, and darkly secret furans.

12

That such a major danger could have gone on for so long with the average citizen left unaware of it—evading the grand tools of environmental engineering, or simply set on the back burner by overburdened chemists—is not just surprising and disconcerting but cause for another acute effect: anxiety.

Hiding behind PCBs, which they virtually always accompany, furans can be found up and down the Mississippi River, whether in Iowa, amid the amber grain, or in the coal-fired, oil-soiled middle of Illinois.

Attesting to their persistence and ubiquity is the fact that they have been tracked in seal fat from Sweden, in snapping turtles from the Hudson River, in carp used to make gefilte fish, in lobsters, in mallard ducks, in cows. They have also been seen in samples of tissue taken from the human population.

In addition to their storage in fat, where they are largely separated from the circulatory and excretory dynamics of the body, the persistence of furans is a tribute to their ability to withstand the power of water, air, light, and other agents in the open environment, which as I said, tend to break many other toxic compounds down. This is a special tribute to the fact that very few strains of microbes—the bacteria and other tiny organisms present in the soil—can digest furans.

I sensed relief among scientists that furans have not caused the grand media splash they deserve. Certainly there was relief among scientists employed by government and industry. They know that a complete perspective on the issue will only lead to widespread and perhaps uproarious public concern over many other compounds, too, and they

115

are certainly not about to initiate such a trend by shifting attention from household names like PCBs and dioxin to a huge new threat like furans.

By 1987, data on furans was still in a primitive stage. The year before, I obtained EPA's first comprehensive review of the existing data, but it was still in draft form with the warning, "Do Not Cite or Quote."

A portrait was quickly taking shape, however. During some years, for example, around 300 million pounds of chlorinated phenols have been produced worldwide as slimicides, fungicides, phenoxy herbicides, and to make a very widely used wood preservative, pentachlorophenol. This quantity and pervasiveness of potential furan carriers should be enough of an incentive for government to begin an emergency investigation of the furan effects.

Certain furan isomers have been attacked in a heavily employed industrial ingredient called hexachlorobenzene, which is a contaminant in the production of many other compounds, including chlorine, vinyl chloride, carbon tetrachloride, trichloroethylene, PBBs (polybrominated biphenyls), and atrazine, the popular triazine herbicide that farmers all over Iowa and the rest of the farm states are using.

As for PCBs, which are now banned from production but still sit out there in electrical capacitors and transformers all around America, they carry furans as another one of those unexpected (and wholly unwanted) impurities that come about during the manufacturing process.

That is bracing news, considering that more than a billion pounds of PCB fluids are in current use, and millions of pounds have already escaped into our air, soil, and water, making them every bit as pervasive as DDT.

Never should one ignore the fact that additional furans are created when PCBs and other chlorinated materials undergo heat. Since PCBs are in use as dielectric coolants—their very job bringing them to warm situations—and since they also have been used as an ingredient in printing materials, which, like other trash, is eventually burned, it is not difficult to imagine where the necessary heat might come from.

I found one journal article estimating that between 1968 and 1975, up to ninety-four thousand grams of furans may have been generated in America as a result of the production of certain kinds of PCBs.

The estimation included enough 2,3,7,8-tetrachloro furans and enough also of a similarly potent isomer, 2,3,4,7,8-penta furans, to kill at least 4.4 to 8 billion guinea pigs.

While scientists often stay clear of such calculations, knowing how very risky it is to extrapolate in this fashion and wishing to avoid anything that smacks of a scare tactic, the numbers, in the case of furans, are misleading only in the sense of being understated, for they do not take into account every type of PCB, and the figures also ignore chlorophenols, PBBs, and other potential furan-carriers that can be found throughout the Midwest.

Furthermore, the tetra and penta isomers mentioned above are only 2 of 135 furan isomers—an additional 10 or so of which are known to be "significantly" toxic to laboratory animals.

Such a wide range of sources means one thing: furans in all probability are much more abundant in the environment than TCDD. They are probably the most troubling environmental poison of all time, with lethal doses that, like TCDD, must be measured in micrograms.

Furans head for the liver, which is as vital as an organ can be, and to make matters more troubling still, this class of compounds seems to diffuse rather easily through many kinds of cell membranes, giving them direct access to a cell's control center.

When they are excreted it is often through a woman's rich child-nurturing milk.

Because they cross the placenta, however, an infant might well be exposed even before nursing.

I drove through Iowa corn country with the arcana of chemistry—like a cumulonimbus thunderhead—hanging dark and heavy.

I could see why scientists were leery of another scare. Starting with DDT and thalidomide and the fear of Red Dye #2, we have been overburdened with "deadly" this and "toxic" that to the point where there is hardly the mental room left for remembering our Social Security numbers.

And the chemistry is like trying to comprehend all the stars in a distant galaxy. No one has actually seen a toxic molecule, at least not the ones we're talking about, but each of the atoms composing them can be pictured as a miniature solar system.

The nucleus is like a tiny sun, with the electrons whizzing around it like planets. Each solar system, of course, is either a carbon or a chlorine or an atom such as hydrogen.

But that's much too easy: Sometimes there is bromine instead of chlorine, sometimes there are double bonds. Sometimes a molecule is flat, like TCDD, or slightly bent, as furans are.

117

All this is of consequence both in the environment and in the body.

If it is confusing enough trying to count all the chlorines, it is worse keeping track of just where they latch onto the benzoid rings. And yet it makes a big difference toxicologically. The effect on the body not only depends upon a molecule's number of atoms but on how far they are spaced from each other. To realize how exquisitely important a seemingly minor difference can be, consider that, in TCDD isomers, moving a chlorine from the "1" position to the "2" position can increase the toxicity between one thousand and ten thousand times over.

What little we know of furans comes from two incidents in the Orient, one in southern Japan during 1968, one eleven years later that started in central Taiwan. In both cases hundreds were sickened by eating rice oil contaminated by PCBs. There was death from liver disease, and a blackening of toenails that placed one in mind of Carol Jean Kruger in Hemlock, Michigan.

In both countries the mysterious disease grew to legendary proportions. In Japan it was known as "Yusho," in Taiwan as "Yu-Cheng." By the end of 1982 authorities had counted 1,788 victims in the Yusho outbreak, which was caused when PCBs leaked from a heat exchanger and into the oil during the manufacturing process.

But in a startling twist to the Yusho case, scientists were soon to decide that many of the ailments were not due to PCBs but to the hidden presence of dibenzofurans.

When chemists and biologists picked apart the rice oil and isolated the various toxic constituents (including another we can add to the lexicon, "PCQs"), they found that administration of furan-free PCBs to monkeys did not cause the Yusho symptoms.

No one doubts that PCBs possess a certain toxic power of their own, including the likely role of a cancer promoter. But laboratory results indicate that furans were the primary agents of harm in what were at first thought to be PCB outbreaks.

"It now appears," notes William Lowrance, a professor of life sciences at The Rockefeller University in New York, "that a lot and maybe most of the toxic action of PCB materials can be attributed to impurities or minor components such as dibenzofurans."

Spooky indeed! All these years, all the headlines, all the notoriety given to PCBs in the Hudson River, and in the Mississippi too (near

118

Davenport, Iowa), now seem to have been a case of barking up the wrong tree.

Or at least of perching on the wrong branch.

All along, it appears, a contaminant that could be described as the structural analogue of dioxin, a contaminant the vast majority of Americans have never heard mentioned—a kissing cousin of TCDD, capable of causing the same chloracne, the same strange body wastage—was silently behind the PCB scene.

In many cases of PCB contamination, when public officials automatically assured us that water or food supplies were safe because the PCB concentrations were not at dangerous human-effect (or "threshold") levels, the accompanying furans had not been taken into full account! There existed the distinct chance that, in secret, they were causing cancers across the continent.

13

Every minute we are taking more than a dozen breaths, every breath a half liter or so, the children breathing four or five times more per body weight than adults, the billions of various molecules heading for the lymph system, the blood, the digestive tract, fighting their way toward trillions of human cells.

And in the pristine, rural air of midwestern America lurks a cornucopia alright—of toxic chemicals.

I was digging into both the Corn Belt and Bible Belt; in Dubuque, Iowa, between lush patchworks of soybeans and corn, the Holy Ghost Credit Union is across Central Avenue from a beautiful brick church.

Further south, in the city of Clinton, I encountered three chemical plants, a power company, and the sweet smell not of volatile solvents but of corn processing.

Nearby, at the Smith Brothers General Store, there were woodstoves for sale, should anyone have been interested in lighting one up to test for furans or benzo-(a)-pyrene.

In Dubuque and Waterloo they painted farm equipment, and that meant solvents like toluene. In Des Moines there might have been some trichloroethylene floating around, since it had been tracked in the water there.

From the west had come some lead from Omaha, Nebraska, where there are also paper mills. According to a government review, the processes used in refining paper sometimes include slimicides and fungicides that may carry various isomers of dioxin.

Fremont, thirty miles northeast of Omaha, manufactures atrazine, and in Lincoln a printing plant had been cited for hydrocarbons.

121

Down in Kansas, chlorine and other caustics are made in Wichita, leading to health complaints, and paint is made in Topeka. In the case of Wichita, there are reports of seizures in horses and pigs with leukemia. Kansas City has a car assembly factory, and in large stretches of the southeastern end of the state, abandoned mines have sent zinc and lead into the local air, cutting lung tissue and perhaps responsible for the area's high cancer rate.

The Midwest uses coal-fired boilers to process sugar beets, and in Illinois, Iowa's eastern neighbor, there are ten thousand industries capable of emitting air pollutants, roughly two thousand of them significant enough to warrant regular inspections by the state's regulatory agency.

The chemicals sweep off the hilly fields in gusts and dust devils, or evaporate from tank solutions, unnoticed by the eyes or nose.

One Iowa official told me the windblown dust has not been tested but is probably "somewhat of a concern."

Perhaps even a very major one: Besides all the triazines and 2,4-D, which have already given federal scientists more than they can evaluate, there is the fact that to further cut costs many herbicides, insecticides, and fungicides are mixed and applied together in one operation, opening the question of how they react together and whether any new compounds are formed in the process.

One thing we do know is that under certain atmospheric conditions parathion may be converted into a compound which is ten times as toxic.

What a web we weave: The herbicides can change the composition of a field so dramatically that other types of weeds—no longer feeling the competition from those the herbicides have wiped out—grab the opportunity to move in, necessitating the use of a different herbicide or still another tank-mixture.

It is a hit-or-miss endeavor in the truest sense of that term. According to one popular estimate, less than one percent of the poisons reach their target pests.

A disconcerting amount of the rest finds its way into the human body, is metabolized and distributed through the blood stream, attacks the tissues and their cells, and distorts DNA molecules, producing mistakes that allow cells either to proliferate wildly or trigger the replication of cells that are grossly altered.

Whatever the role of twisted molecules and hidden poisons, it has been learned recently that farmers handling weed killers face a far higher risk of non-Hodgkin's lymphoma. If they are exposed to them twenty days or more a year, they are 600 percent more prone to the serious illness than the general populace.

Among the hundreds of active pesticidal ingredients, the EPA, by early 1986, could confidently offer safety assurances on only thirty-seven. A number of pesticides registered for use during the past three decades have since been found to cause birth defects and cancers. Time and again, however, the federal government has waited until the human toll adds up. Guinea pigs, it seems, have been sacrificed for naught: The compound ethylene dibromide was not banned as a pesticide until some ten years after hints that it was carcinogenic. In other cases, test results that claim certain chemicals cause no adverse effects have been found to be fraudulent.

In Iowa, one recent year, more acreage of corn was treated with herbicides than in Michigan, Minnesota, and Wisconsin combined.

The bigger the farm, the more pesticides; the more pesticides, the lower the costs. So, the bottom line is once more the only one we read: Whether in the emissions of a factory stack or out here in the spraying of corn, the chemical molecule always revolves around the dollar sign.

Along the routes I followed were roadside shanties dedicated solely to the distribution of agrochemicals. Actually the word *shanties* evokes the wrong image, since these shacks are shaped in shiny new aluminum shells—more prosperous in looks, to be sure, than the surrounding barns and silos.

The products are touted in glossy, happy-go-lucky pamphlets. Inside are photographs of fresh, yearning seedlings—as if the chemical was their elixir of life. On one cover the herbicidal mixture looks like a plastic jug of cherry Kool-Aid.

The actual practice of application is more unsettling. Sickly green liquids composed of eight or more separate products are sprayed directly into the loamy dirt. Or mixed in pivot sprinklers with the very irrigation water.

Their immediate aim is of course not to nurture life but to interrupt the very foundation of it, stopping an unwanted plant's photosynthesis or otherwise causing it to burn out. That is how a weed is killed.

Soon, I was told, eighty or so additional herbicides will be added to the list. If trained scientists have trouble following every one of them, I wondered, what about the farmer who has to read all the complex labels?

When I snuck onto a farm near Durango to inspect the metal barrels, I noticed one containing paraquat, the herbicide that created a hot controversy when it was used to kill foreign marijuana crops. Farmers were having particular problems with black cutworms that year, and some of the other pests—pigweed, ragweed, smartweed—sounded like players in an old-timey cartoon.

To give the insecticide advocates their due, it should be noted that without modern chemicals virtually all apples would have moth stings in them, ears of corn would be stained with larvae excrement, and if there were rootworms around, the cornstalks would bend in such a way as to inhibit mechanical harvesting.

One technology holds hands with another, and in the end there is the devil to pay.

"There are more questions than there are answers," said a state environmental specialist, Cindy Cameron Combs, when I asked about the potential of the new pesticides for ecological effects.

"We had thought they would break down within the farming season, but what we're finding is that the rate of decay may be different in river-bottom sediments. Where before we thought, 'They'll break down. We don't have to test for them,' we started seeing them in shallow wells. They're longer-lived than we originally thought."

What happens when they mix?

"Nobody knows anything about that."

We do know, however, that in combination, one chemical may destroy the liver enzyme which normally would render a second, accompanying compound harmless. In the instance of certain mixed chlorinateds, the toxicity can thus be increased fifty or a hundred times over.

Or, in theory, one chemical may act as a tumor initiator, another as a promoter. Such toxic teamwork has only begun to be investigated, but the result may already be available in cancer statistics. Dr. Michael R. Greenberg, who is perhaps the nation's foremost authority on urbanization and cancer rates (and who, as a resident of New Jersey, is certainly not sitting in an ivory tower), told me his most recent data shows that the once wide gap between cancer mortality rates in big cities, where they were highest, and in rural areas, where they were traditionally lowest, is closing, for whatever reasons, "very, very rapidly."

Indeed, the poisonous nature of modern rural America hits home when one goes beyond the Kool-Aid brochures and realizes that certain pesticidal agents bear a chemical resemblance to what is found in the fly amanita, a poisonous mushroom. Or when one looks at the cellular effects they cause. In the case of 2,4-D, these effects parallel those caused by an x-ray.

There are also residues of hexachlorobenzene to consider, used as it was as a fungicide (until recently, when such use was halted because of its hazards), and pentachlorophenol, the wood preservative that can be found on rail ties, wood posts, or in silos themselves—alongside the herds and feed. In 1980 annual production was estimated to be 53 million pounds, which meant significant dioxin contributions of the hepta, hexa, and octa types, as well as furans when the treated wood is burned. Though it would appear to be highly unlikely, and would presumably involve irregular circumstances playing off its phenyl content, one federal report mentioned that dioxins might also be associated with the major insecticide parathion.

Added to the thousands of active pesticidal ingredients in use are the hundreds of inactive (or "inert") ingredients which are employed in a product to dilute, propel, stabilize, or otherwise enhance the effectiveness, shelf life, and thus salability of the pest killer. According to a report by the General Accounting Office, twelve hundred to thirteen hundred chemicals are registered for this secondary role, and while a good number are innocuous—so harmless in theory that they can be used in foodstuffs—there is not enough information to evaluate the toxicological implications of more than eight hundred of them.

It *was* known, however, that at least fifty-five are of immediate concern, half of these potentially carcinogenic. Among the inert pesticidal ingredients on an EPA list were asbestos, benzene, carbon tetrachloride, cadmium, pentachlorophenol, and hexachlorophene, yet another possible dioxin carrier.

To add a little more to our worry quotient, I also came across a paper from the National Institute for Occupational Safety and Health (NIOSH) that barely mentioned the potential problem of windblown, chemically contaminated soil because it was concerned with dust from the grain itself. This can elicit acute and chronic respiratory problems and introduce to the body exotic toxic substances that come from fungi and bacteria, such as endotoxins, mycotoxins, and alfatoxin—none of them good for the body's all-important immune system.

One last note before we look beyond the soybean patch: A recent study directed by Dr. Eldon P. Savage of Colorado State University has seemingly shown that aside from the maladies commonly associated with pesticides, the organophosphates can also affect our very psychology.

This type of research is especially tricky but worth consideration nonetheless. In looking at one hundred agricultural workers who had been exposed to poisonous doses between 1950 and 1976 (versus one hundred who had not), the researchers noted a significantly higher incidence of depression, irritability, anxiety, and paranoia. The victims also seemed more socially withdrawn and performed worse at academic and motor skills. "The differences were quite marked," Dr. Savage said.

Down in southeastern Iowa is the spired and turreted city of Muscatine, with a Monsanto herbicide facility that manufactures Lasso, the brand name for the suspected carcinogen alachlor. The nation's largest chemical firm, Du Pont, appears with a small plant in Fort Madison, making thinners, resins, and paint.

Before the Missouri border I spotted pink lagoons of wastewater.

And soon, from the South, came a steady, rumbling stream of tank trucks.

We are in the middle of the Superwhip, where a little community had been turned into a bona fide ghost town, and where little tragedies threatened to become big ones.

14

Down a forested, winding road near the banks of the Meramec River in Missouri is a police trailer at the bridge that serves as an entrance for the small, dilapidated, abandoned town. Next to it is a large yellow sign: "DIOXIN CONTAMINATION. Stay in car, minimize travel, keep windows closed."

Beyond the barricade are boarded homes indeed haunting to behold, their paint peeling, their lawns overcome by weeds and brush. Mattresses and gutted appliances are piled here and there, the valuable stuff looted already.

No longer does the community exist. It has been relegated to history, taken off the map—a disincorporated twilight zone.

Severe spring thunderstorms again advanced across the horizon, their boomers loud in the hills near this little floodplain town.

There was a time when Times Beach was home to 2,242 citizens, four churches, a ballpark, and thirteen taverns. That was long before the Meramec, a tributary of the Mississippi, overflowed its banks and deluged the community, which is about seventeen miles southwest of St. Louis.

It was also before authorities found dioxin in thousands of parts per trillion, and, as at Love Canal, evacuated the place.

In the same way that St. Louis and its arch are the Gateway to the West, Times Beach and its soil are now the symbolic gateway, in the sense of toxic contamination, to the great Southwest. The town is but one of forty-five sites in Missouri recognized as having a problem with TCDD, which had been mixed with waste oil and sprayed for dust control. The flood helped to spread the dioxin around Times Beach, but most of the credit for putting it there belongs to Russell Bliss, a waste hauler and talented enough a double-dipper to have qualified for the East Chicago City Council: He had been paid to discard the chemical waste, according to state officials, and then he collected another fee to spread it on gravel roads.

Incredible as it might sound, dioxin had been sprayed, dumped, or used as grading fill near a home for the aged, near a Methodist church, at calf farms, and in horse arenas, where birds literally fell from their roosts and horses died in droves.

The solvents benzene, xylene, and toluene had also been found in substantial quantities on a Times Beach baseball diamond.

In the meantime, at the other side of the state, it was suspected that a Kansas City dumping operation had created furans by burning PCBs. The waste site, now inactive, was on the banks of the Missouri River this time. According to a state report, "a number of areas are barren of vegetation with no apparent cause."

Though waste disposal is not the subject of this book, such details are offered as clues to new airborne dangers that are taking hold everywhere. Each year 550 millon gallons of waste oil are recycled as fuel in the United States, some of it burned in kilns, most in boilers.

It is highly laudable to reuse a valuable and finite resource, but not if poisons have been mixed into it. Though TCDD is not a common ingredient in recycled petroleum, arsenic, cadmium, benzo-(a)-pyrene, and PCBs—furans—typically are. To one degree or another, they will escape with the flue gas.

One reason Times Beach rated as a crisis is because dioxin, like the farm pesticides, can hitch a ride on those airborne particulates or volatize out of the soil like a vapor if mixed with the right solvents. While Americans view landfills and other waste sites chiefly as a threat to groundwater, and thus to our wells, there are many times, as in south Chicago and now near St. Louis, that our most immediate danger is not below our feet but over our heads.

When something is in our ambient air, it is probably more of a threat than anywhere else. "It is likely that the total population dose is much greater from air pollution than any other media contamination," says Anthony D. Cortese, director of the Tufts University Center for

Environmental Management. "Humans drink at most two liters of water a day. In contrast, we breathe fifteen to twenty-five thousand liters of air a day. And the functon of the respiratory system is to exchange gases with the blood. Even assuming a differential absorption rate between inhalation and ingestion, air pollution exposure dwarfs exposure from drinking water."

And it spreads much farther. In Louisiana, hexachlorobenzene once vaporized out of waste pits and contaminated an estimated one hundred square miles of pastureland, along with thirty thousand head of beef. If such dramatic volatization and subsequent fallout can occur in Louisiana, then surely it must also take place, to a degree, at some, if not most, if not *all,* of the other hazardous-waste sites that contain volatile organics and are likewise unraveling out of technological control.

In New England and New Jersey, very high levels of contamination have been detected above such wastes, to the point of clogging the test equipment. In California the waste sites in one community collectively pump up at least as many fumes as a major refinery.

It is another area where the numbers are numbing. Two widely used figures: there are 10,000 hazardous-waste sites currently posing a serious public health threat and a grand total of 378,000 that eventually will need corrective action.

Whatever the merits of such high estimations, during 1981, the most recent year for which the EPA's Office of Solid Waste had comprehensive figures, America generated 581,988 million pounds of wastes which were officially designated as "hazardous"—more than a ton per person. Of that quantity, the majority was stored above ground while awaiting treatment, recycling, or ultimate disposal, and hence was exposed, in some measure, to our air.

America the Disposable! On industrial back lots across the country, waste liquids are leaking out of rusted barrels or evaporating from expansive, multicolored, steaming lagoons.

Or escaping from sludge burners.

In Missouri, health officials felt there were no clear-cut, clinical illnesses that had resulted from the dioxin exposure. With a compound that kills certain laboratory animals in doses of less than a microgram per kilogram, one may have expected actual carnage at Times Beach: at least a case or two of people strolling near a hot spot and instantly keeling over.

Or massive weight loss and chloracne.

Perhaps even a town of people who looked like the living dead.

Instead, there were the Kleins, an older couple who had refused to leave their home.

Long before, the EPA had announced that it was buying out and closing up the town, but the Kleins were dissatisfied with how much the government offered for their property.

They had lived there forty-four years, and though Mr. Klein had had colon cancer, according to his wife, Lorene, they had suffered no illness that could be related to the dioxin and felt the press "had blown it all out of proportion. They made it sound so contaminated and everybody ran."

Now they lived among the spoils.

Existing on Social Security payments, the Kleins stretched their dollars by relying upon a large garden for most of their vegetables—and for dessert treats such as raspberries and strawberries as well.

"It hasn't affected us," said Mrs. Klein, standing near the garden. "And the animals, *they're* not bothered by it. I haven't seen any dead cats. And you see a lot of birds and rabbits come through here."

Once Times Beach ceased to exist, she and her husband had to haul their garbage to a daughter's house in another, functioning, town.

She smirked when I asked how it was to live in a ghost town. "Beautiful. We don't have anybody to bother us."

But then she turned serious. "Where are we to go?"

By the autumn of 1986, three years after the evacuation, the government had forced the Kleins to find another place to live, and they were finally in the process of doing just that (having a problem finding movers, however, who were willing to come into the contaminated town).

"We really didn't want to give up our home. The kids were all raised there and the grandkids were raised there. And, ya know, it's been kind of rough on me, it's added quite a few years to my life," Lorene said flatly, as if, by speaking without overt emotion, her homesickness might be dispelled.

The dioxin was measured at 300,000 parts per trillion on some roads at Times Beach, according to my guide here, Don Maddox, regional administrator for the Missouri Department of Natural Resources (DNR).

High levels were found at some ten or so other sites where residents were offered relocation, including a trailer park to the west called Quail Run and a residential area in the adjacent county where the TCDD-type dioxin was detected on household dust.

130

Taking into account a dazzling array of simulation models and log-linear symbols and equations that mean nothing to the layman (for example, $TCDDt = e [a - 0.00016 \times t]$, which stood for soil levels and was one of the simpler ones), the CDC in Atlanta determined the threshold of concern, when dioxin is in the soil, to be one thousand parts per trillion—$\frac{1}{300}$ the amount in spots at Times Beach. The figure was conjured by estimating the various routes of exposure (including kids playing in the dirt) and what exposures would take the residents beyond a "reasonable risk"—the reasonable risk defined as about one excess cancer per million people, according to Renate Kimbrough of the CDC.

This is known as "risk assessment," and it is at best a dubious exercise. In most cases it involves assumptions built upon other assumptions, the effect comparable to a house built in sand.

No matter how impressively complex and exotically symboled, no system of mathematics can really factor in every conceivable variable and predict all the effects compounds will have on every enzyme and tissue in the body. There is no way of precisely knowing how much has entered a person's body and how long the toxicants will persist there. As Dr. Kimbrough sums it up, "There are always uncertainties."

Dr. Bernard D. Goldstein of Rutgers Medical School once wrote that "the concept of a threshold implies that there is some dose of a compound which is not capable of producing the adverse response. It should be noted that we never really determine a true threshold; rather it's possible to plot a highest no-effect level determined in a study."

Chemical levels previously not thought to be a problem at tiny levels in the blood have proven otherwise in certain people. Mild chemical hepatitis—liver damage—has been reported in a woman whose blood level of the chlorinated pesticide heptachlor epoxide never exceeded three parts per billion. I once met a woman in Buffalo who, in an allergic reaction, lost her voice if she was merely around plastic materials.

Nor, again, is there any way of knowing how different chemicals will react with each other. We do know, however, that in some instances dioxin has been found to enhance the action of other compounds. Rats treated with sublethal quantities of TCDD prior to administration of the anesthetic hexabarbitone slept twice as long as they would have without TCDD.

This may be due to altered enzyme activity induced by the dioxin. Acting ever so subtly, TCDD seems to suppress various mechanisms of the body's defense system, our immunological functions, opening us to infectious assaults that will probably not be properly recognized, by the

131

average physician, as a chemical effect. Mice have died quickly and in increased numbers from *salmonella* infection after they have been exposed to slight quantities of TCDD.

It can also promote the cancer-causing potency of other molecules —a particularly worrisome sidelight, since dioxin by itself is already considered to be so very carcinogenic. In fact, its major role in cancer may be that of a tumor promoter as opposed to being a classic carcinogen in and of itself.

At Missouri's dioxin sites there was, once more, a huge, baffling gap between the effects claimed by residents and those noted by medical officials. If residents saw a plague of seizures, miscarriages, and hemorrhaging organs, Daryl Roberts of the state health department had "not seen anything that has shown overt clinical illness—no cancer, no chloracne."

Skin responses among residents of Quail Run suggested that TCDD had worked to suppress their immune systems. When exposed to antigens, their systems did not react normally, and possible effects—subclinical effects—were also seen in the results of liver-function tests.

But a previous study had failed to find overt pathological trends— at least none that occurred at a level that was "statistically significant."

"It's a typical example of obfuscation by industry, government, and the press," says a very prominent midwestern research physician whose work in defining the actual mechanisms of dioxin toxicity, including such esoteric details as its binding to rat-liver proteins, is extensively quoted in most major reviews.

"[Dioxin] has been overemphasized by the press and government. It's been carried to an extreme. It's been extraordinarily and irresponsibly exaggerated, and some of that is by scientists: If you create an imminent health hazard, they're likely to continue your funding.

"Newspapermen are fire chasers, and they go [on] to the next poison, and they don't look anymore for a guy's credentials. A lot of industry people are surreptitious, and government people are dumb."

On the other hand, in the same technical reviews that featured this scientist's work, I noticed, TCDD was described as a carcinogen nonpareil. It is, said one study, fifty times more potent than the third-ranked carcinogen, bis(chloromethyl) ether—and thirty thousand times more potent than a chemical called 3-MC whose very value in cancer research has been its quality as a cancer-inducing agent.

It is very hard to embellish such facts. More than anything, at any rate, dioxin stands as a metaphor for the other chlorinated compounds of our age. It is a tracer, a harbinger, a thin swipe of angry red paint on the larger toxic canvas.

There is reason to believe that dioxin and furans are also in places such as St. Joseph, Verona, Springfield, Sturgeon (where there'd been a serious rail spill), and across the western border in Kansas City, Kansas. It is also a reasonable bet that they are across St. Louis's eastern border in Sauget, Illinois, home to the Monsanto Company, which had invented the very PCB molecule in the first place and had been the sole producer of it in America.

Some of these toxicants are no doubt swept up in the thunderstorms to commingle with the other exhalations of the St. Louis area. The city itself had a very high annual cancer rate—in white males about 265 per 100,000—but St. Louis is not as high as Chicago in toxic, ozone-forming vapors such as benzene, xylene, and toluene.

According to one set of samples, St. Louis's toluene was less than a sixth the levels seen in Houston and Los Angeles.

But its carbon tetrachloride has been high, and so have been the emissions of methylene chloride—higher, certainly, than a similarly old-school and box-factoried city like Pittsburgh.

The state said that a nearby Dow plant emitted 12,000 pounds a year of acrylonitrile (which is a nice way of saying "vinyl cyanide"), while Monsanto was said to release 337,600 pounds of chlorobenzene, an annual amount that Harold J. Corbett, senior vice-president of the company, says does not sound atypical. "When you're dealing with millions of pounds of chemicals, it's entirely possible that one thousand pounds a day can be emitted of any chemical," he said, adding, however, that the concentrations in the actual open environment are very low.

According to the state DNR, Hercules Chemical once released 28,800 yearly pounds of formaldehyde in another Missouri town that is ironically named "Louisiana."

Another source of air pollution in the Midwest is sewage plants. At least fifty-nine of the chemicals classified as "priority pollutants" by the federal government are in the river water and effluents of the St. Louis vicinity. Their evaporation from treatment cells—and from the Mississippi itself—has to count for something.

133

There are days, according to my calculations, when as many as 20,929 pounds of trichlorobenzene move in the river as it passes St. Louis.

More blatant was the lime-colored smoke I spotted coming out of one stack between St. Louis and Mark Twain's hometown of Hannibal —in the vicinity of that town called Louisiana. Such emissions serve well to enhance whatever pollution Missouri imports from other states. The strongest spike on the wind rose shows the air currents channeling straight up the Superwhip from Arkansas, Texas, and the *state* of Louisiana.

That was an ominous sign, and so was the fact that—speaking of Mark Twain—elevated levels of ozone had been recorded in a national forest, a wilderness, named after the writer in southeastern Missouri.

According to Gary Evans, a chemist with the EPA at Research Triangle park in North Carolina, the agency had determined that much of the ozone over the Mark Twain National Forest is sent up by the massive Baton Rouge–New Orleans petrochemical corridor 450 miles south. It takes about two days for the stuff to get that far north. From there, if wind roses can be believed, it moves in the direction of Chicago and Indianapolis—there to join the windblown dirt from Bloomington that probably carries a few PCBs of its own.

Over central Illinois, on the way, it surely picks up dangerous insecticides. In 1980 a little-known set of government tests found such substances as chlordane, malathion, and diazinon in the air near Peoria.

That was all well and good as far as the general recipe of northeast-bound, Middle-Superwhip pollution, which took the big raw meat from Louisiana and Texas and stewed it in spice and vinegar.

But such sweeping perspectives tend to miss little hot spots of controversy such as Sedalia, Missouri, where thirty-one citizens had sued Alcolac, Inc., which makes shampoo, and were awarded damages of $49.2 million around Christmastime in 1985 in a bizarre, confusing case. Some of the people claimed that bubbles and foam drifted from the plant, and according to their expert witness, Dr. Bertram Carnow of Chicago, "everything had been zapped within three or four miles."

Rashes, fatigue, body sores. Claiming there had been serious damage to their immune systems, Dr. Carnow flamboyantly described it as a case of "chemically induced AIDs."

If the people lived long enough, he fairly shouted about, they were destined for cancer.

There had also been an interesting suit back near the river on the

Illinois side. It was against several large companies for caustic releases which had sent hundreds to the hospitals. This is East St. Louis, and it is just chock-full of grit and soot—a true underside of industry.

Sauget is right next door, and so, once upon a time, was Dead Creek —a dry, smoldering bed of PCBs and phosphorous.

Clouds of hydrochloric acid and other caustic materials have also drifted from barges moving up on the Mississippi, which joins with the Missouri River (a dozen miles above East St. Louis) at a point that had enough dark smog and burning piles of shoreline rubbish to give it the glow of an inferno—a littered, oily, ugly piece of river.

Furans had to have been in the air, for there were whole parcels of land filled with the ash from incinerators.

Back across the river, two of St. Louis's municipal trash burners had been found in 1985 to be in violation of clean air standards.

I spoke to one city woman, forty-one-year-old Else Ann Hirzel, who is hypersensitive to the area's pollution, which shuts down her lungs. Sometimes the asthma is so bad she too must rush for a hospital. Most days she cannot even ride a bus without getting ill.

So she had taken a number of unusual measures to forestall the next crippling attack. "When we are going on a long trip or things are really bad around me, I will wear either a carbon-filter mask or an air helmet that has a battery-powered motor and blower system," she said.

"It sucks in air through a double-filter system.

"With the helmet I feel strange, self-conscious—people stare."

But, she was quick to add, "There are also times I'll take an oxygen tank."

15

If it seems, by now, that everyone might soon need a moonsuit it is time for a bit of perspective.

For by compiling case upon pollution case, we are soon brought to think of everything as one big contaminated mess, which is a grossly inaccurate picture.

Even within badly afflicted areas, pollution levels are not necessarily high everywhere around. The air a few blocks from a contaminated neighborhood can be hundreds of times cleaner than the problem zone.

More to the point, some of the most industrialized states also have large, unspoiled expanses of swamp, forest, desert, mountains, or prairie. We must realize that the largest city in a state may occupy less than one percent of that state's land. Between the industrial amassments that serve as our morbid examples of air problems are usually surprisingly long servings of fresh terrain.

I moved south past Cape Girardeau and into the small end of Kentucky that abuts the Mississippi, by barbecue pits and low thickets draped with flourishing vines.

Still, if you look hard, you will always find some chemicals. Kentucky and the state below it, Tennessee, are both substantial contributors to industry. According to a 1982 census, in fact, Tennessee outranked Missouri and also Michigan in total value of chemical shipments, while Kentucky was ten rungs below Tennessee, in the slot just after West Virginia.

Kentucky had had problems with vinyl chloride in its air, and the broad mix of manufacture near the Ohio River—the plastics, synthetic rubber, the odors of ethanol from whiskey storage—all did a little in pushing up ozone levels in such areas as Louisville and the counties at the far north-central part of the state, just below Cincinnati.

In Jefferson County alone an estimated 50 million pounds of volatile organic compounds were released one recent year.

Both Kentucky and Tennessee were among the states included in a four-year study in which thousands of people were examined for pesticides in their blood and urine by the National Center for Health Statistics and the National Human Monitoring Program of the EPA. Of six thousand urine specimens in one aspect of the study (which took place in sixty-four locations), 79 percent showed levels of pentachlorophenol. Carbofurans (a pesticide), 2,4-D, and trichlorophenol were also detected in the urine.

Virtually everyone tested had some quantity of the chlorinated hydrocarbons in their systems, and Tennessee, especially the Memphis area, had been a hot spot both in the sense of manufacturing chlorinated pesticides and in being contaminated by them. One of the firms there, the Velsicol Chemical Company, had a notorious past as a polluter.

Among this family of chemicals is DDT, the most infamous insecticide of all time. It is a category of chlorinateds that takes us deeper into the realm of side-chains and branches, glued together with bonds that are chemical groups in their own right. Listened to closely, the technical name for DDT proves this out: dichloro-diphenyl-trichloroethane. Phenyl again!

The structure can be pictured in a way that reminds us a little of dioxin and PCBs because there are two benzoid-shaped rings attached to each other.

But with DDT the bond between them is a toxic product unto itself. It is the solvent called trichloroethane, which smells like chloroform and has found use on its own as a dry-cleaning fluid.

Like dioxin, the worry is more their long-term consequences than the type of acute effects associated with parathion and other of the organophosphate pesticides. Chlorinateds can last decades in the environment. Though DDT was banned in the early 1970s, it is still very much with us and *in* us.

In Memphis, according to the health department, dieldrin and heptachlor epoxide, two other powerful, long-lived chlorinateds, have been tracked in the air. Because of their cumulative effect, piling up in our fat, any amount whatsoever is quite unwelcome.

Another one, the termite-killing chlordane, contaminated the Mississippi to the point where eating its fish and those in tributaries had been warned against on riverbank signs. It was accompanied by yet another chlorinated called aldrin, which, in the air, is perhaps thirty-five times as dangerous as benzene. A quantity the size of an aspirin could kill several hundred quail.

Thanks in part to contaminated fish, meat, and vegetables—which have encountered insecticides not just through direct spraying but also by way of toxic atmospheric fallout and wind erosion—96 percent of 785 fatty-tissue specimens surveyed in the National Human Adipose Tissue Monitoring Program of 1978 showed residues of heptachlor epoxide, and 95 percent showed evidence of dieldrin.

As the Memphis health manager put it, the air "can and/or does carry hydrocarbons from long-range sources (i.e., Texas and Oklahoma oil wells and refineries) in addition to who-knows-what."

Far to the southeast, at the opposite end of Tennessee, there is a tract of fifty thousand acres so denuded by past smelter emissions that it has been spotted by orbiting astronauts. Nashville is heavily into printing, Tullahoma has witnessed some yellowish plumes from rocket-engine testing. In Maury County, where pesticides are made, parathion was once tracked in the air.

The coking operations in Chattanooga, lending benzo-(a)-pyrene to this neck of the woods, helped make that city a well-known center of classic air pollution, and in Knoxville there is more tar in the form of asphalt plants.

In Smithville, said Robert Foster, chief of the state's technical services, clumps of asbestos—clumps, not just microscopic fibers—had been found floating through a neighborhood and gathering on porches.

To add some intrigue, there are also the nuclear weapons and reactors of the Oak Ridge National Laboratories, which may emit radionuclides and chemical cooling agents. "We don't know, totally, what's going on," confided one state official who did not want to be named.

Back in Memphis I learned of the spills and explosions which have occurred with uncomfortable regularity there—one involving a compound closely related to the one released in Bhopal. Workers at the sewer plant have been known to sicken from the fumes, and even explosives such as TNT have been dumped with abandon.

But it was just across the Mississippi, in the air flow from the west, in Arkansas, that surprises were in store. Under the bridge leading from Memphis was a shanty raised above the floodplain on high stilts, near radio towers.

The shack was inhabited by a hermit named William Roberge, who looked after the transmitter for a local gospel station.

He also found time for his fifty cats and his folk guitar. William was a shy, modern-day Noah, collecting the animals when the river crested high and making sure they had a dry place to sleep. Sometimes, to get to his shanty, he had to use a little boat.

But what upset him far more than the intrusions of nature was a little chemical company located just above the riverbank. At one time a brilliant orange chemical was piped directly into the river from it, and his cats wandered into the crud, discoloring their fur.

"A couple of 'em came back goofy—they couldn't stand up," he said. "I sent a sample to the state but never did hear back."

Though the pollution was not as bad as once it had been, Roberge occasionally saw yellow clouds rise from the site, and he sniffed sulfur in the breeze. I tried to find the company's president but was refused admittance to the plant's offices.

This, however, was but a mere prelude to what was going on elsewhere in Arkansas—a state one would expect to be *all* huge expanses and fresh prairie terrain. A sunset that splashed marvelous striations of magenta and apricot through another severe, clouding sky had the elements of a metaphor, for Arkansas yields a set of dioxin circumstances which is more interesting, in some ways, than Times Beach.

16

Just to the northeast of Little Rock, on the flat terrain where soybeans and rice are grown, is the city of Jacksonville. Here, as in Midland, there was manufacture of Agent Orange. The 2,4-D and 2,4,5-trichlorophenol had been made first by Hercules and then an outfit called Transvaal, a subsidiary of the somewhat mysterious Vertac Chemical, which had corporate offices in Memphis but whose ownership no one seemed quite sure of.

Jacksonville is a small city—only 27,589, according to the last census—and the neighborhood around the site is both white and blue collar. It includes retired military personnel who, if they hadn't been exposed to Agent Orange during Vietnam, now had that opportunity through the local environment.

As far back as 1965 farmers and sportsmen had complained that fish were being killed by chemicals flowing down Rocky Branch Creek into the Bayou Meto—to the extent that the fish couldn't be sold because of the offensive smell. In some spots, minnows that were dropped into the contaminated water died almost on contact, and later, because of continued chemical releases in the area, residents were to complain of a medicinal taste in garden tomatoes, along with squash that grew crazily and spots where the grass had turned red.

Leaves curled, as if dosed with an herbicide, and old, sturdy trees suddenly died. Something in the air also seemed to be eating at the driveway cement.

According to one former employee, Jack Park, an analytical chemist who had served as technical assistant to the manager of the plant until about 1970 (and lived nearby), gas from the process had been vented "every batch, which would be every two or three hours," and there had been instances when the rupture disk on the process vessel would start whistling and set he and his co-workers running for cover.

In an explosion he said occurred in the mid-1960s and was not unlike the accident in Seveso, Italy, the force was so great that it blew the loading rim off the apparatus, and the dioxin-contaminated materials spewed out a stainless steel pipe until about a ton of the stuff had escaped outside.

The smell was like licorice: "I'll never forget it—the strongest licorice I've ever smelled."

Afterward, Park said, he developed a skin disorder.

Park's testimony came from the standpoint of a born-again advocate. By now he had turned into an unabashed, quip-prone environmentalist, and officials had no documents showing that there had been a truly catastrophic release.

But, acknowledged Dr. Phyllis Garnett, director of the state's Department of Pollution Control and Ecology, there were records which demonstrated that liquid material sprayed through the stack. "It would spew out chemicals—the real stuff. It was like the lid would come off. During the years of production of 2,4,5-T— in those years everything I've read is that there were the most deathly emissions."

Garnett said a 1980 analysis at the now-closed site also had picked up dioxin particles blowing through the air! But when I repeated back to her what she had told me, Garnett changed the adjective "deathly" to "toxic emissions that would have posed a health risk."

Dioxin had been found in some residential yards near the contaminated creek, and according to site records from 1977, it was not uncommon for chlorophenols to find their way into the air.

Phenolic materials used in herbicidal production could be awfully fatal. In 1985 a Vertac worker named James E. Cox, thirty-two, died after being splashed with about three gallons of such ingredients, and when the paramedics got there they found him scalded and writhing in a shower stall.

So contaminated was Cox that one of the paramedics claimed to have taken ill from inserting a breathing tube into the dying man, whose last exhalations burned the paramedic's eyes.

Park, sixty-six, was more concerned about what had gone on in the neighborhood itself, where a couple of other plants also handled toxic materials. He had suffered a stroke nine years ago and blamed that on the air pollution—the red grass, he claimed, was from chemicals forming a dye in the air.

It ate the spring right out of a door lock, he said, mutated the vegetation. And what about the birds? The creatures would fly through and die on the spot—littering the backyards.

"They'd stand there on one leg like a stork and rock back and forth a couple times and then just fall over," said the former corporate chemist.

Another resident, Patty Kirkpatrick Frase, who had since moved out of polluted Jacksonville, also remembered releases at the herbicide facility. "In 1974 they had a blowout during their nighttime production where the reactor vessel caught on fire and blew out, and it destroyed everything around that area. There were always dead animals out there, even now. All the oaks, all the hardwoods in that area practically disintegrated overnight. We had dead squirrels, dead dogs, all the fish died in the discharge area in Rocky Branch Creek. I mean, that's pretty rough when even the algae in the area dies."

When I asked if she had noticed any human ailments afterward, Patty answered quickly: "Yeah—that's when my father died. He had just retired from the Air Force. They did an autopsy and they really couldn't find anything and they decided his colon bursted.

"And my mother died right after that of liver cancer. There's a lot of liver cancer in the whole area. My girlfriend's mother died and she was younger than my mother—I think she was forty-three, she died of cancer a few weeks later, cancer of the bone marrow and liver, and it was just all over her body. And the man that got me involved in researching Bayou Meto and Vertac was a duck hunter named Rex Hancock who's a wonderful man—he *was* a wonderful man. He just died a month ago of liver cancer.

"And I had five miscarriages. I've been through everything to have a baby—I've been to the moon and back."

Particularly frightening was the fact that there were wastes sitting over at Vertac that even the state of Louisiana had refused to accept for disposal. Patty, whose husband was an advertising executive (and who finally had become a proud father ten months before our conversation), began checking a part of the local cemetery known as "Babyland" to keep track of how many youngsters were dying and of what diseases.

Babyland was where tiny infants who were not buried in family plots could be interred free of charge.

143

"We've had one baby born with its heart outside its chest cavity, and we've had a bunch of blind babies born around the area and we've had many other babies without part of their limbs and other deformities born around the area," she informed me, an echo of Hazel Johnson's voice in Chicago.

"I went deaf in my right ear while I was out there, along with other people. We can't prove any of this stuff, but one thing that's really funny is that four of my friends in Jacksonville and myself all had swollen adenoids under our armpits and on our bodies and we were all told we were allergic to deodorant and as soon as all of us moved out of Jacksonville our adenoids were all normal and we were no longer allergic to deodorants."

Getting back to Babyland, one of those buried there was three-month-old Joseph Shelton, robust and gregarious right up until his unexplained end.

He died on September 6, 1985, of what was labeled as Sudden Infant Death Syndrome—a week after he'd learned to laugh.

"See, we was sleeping, and my older boy, he kicked me, we were all sleeping in the same bed, and he kicked me and that's when I found the baby dead," Mrs. Shelton, nineteen, explained in a subdued drawl, adding that the child had seemed healthy hours before when her husband had fed him.

What made his death all the more remarkable was that on December 18, 1985, the young, impoverished couple were sent a letter from Arkansas Children's Hospital informing them that tests of the baby's autopsy tissues showed that chlorophenols had accumulated in his liver and kidneys at up to 508,000 parts per trillion—in the same range as the dirt at Times Beach.

The dead infant's brother, seventeen-month-old Jeff, began having seizures a week after Joseph died.

Dioxin-related phenols were also found in *his* body, and in Mrs. Shelton as well. The episode cast quite challenging doubt on the oft-quoted claim, by industrialists, that dioxin has never caused a human death.

But the doctors in Arkansas were unable or unwilling to connect chemicals to either the older boy's seizures or the baby's death.

Contamination of children has grown into an inexcusable and extraordinarily despicable modern trend. Years ago tests by the EPA detected DDT and PCBs in 100 percent of the milk taken from fifty-seven lactating mothers in Arkansas and Mississippi. Also present were chlordane (45.6 percent) and dieldrin (28.1 percent).

That PCBs were present infers, in short, that a new generation of Southerners is getting its first body-forming nourishment from milk tainted with furans.

Nor is this simply a problem from the past. Another potential source of dioxins and furans could be found 120 miles south of Jacksonville, in El Dorado, where there is a commercial incinerator taking in PCBs that wanted to increase its burn rate to 5,210 pounds an hour.

The people, some of whom lived within a couple thousand feet of it, were in an uproar. It was so hot and heavy that Vice-President George Bush, swinging on a fund-raiser through the state, was drawn into the fray. The prominent and poor alike had marched in the streets and jammed auditoriums. The incinerator is owned by a firm called Environmental Systems (or ENSCO).

Its location so near a populated area was doubly unnerving to the residents when they realized that elsewhere, public outcry had delayed proposals to burn similar wastes on an incinerator ship more than a hundred miles out to sea. "From what ENSCO has told us, this incinerator burns more PCBs than any other incinerator in the country, over half of all the PCBs that are found in capacitors," fretted local attorney and concerned citizen John H. Vestal, who had compiled so much material on the site it seemed at times as if he were preparing for a fight to the Supreme Court.

He pointed out that furans had been detected in stack gas during a test sample (albeit in minuscule amounts), and added that the incinerator's plume had a dark tinge at times that indicated unburned hydrocarbons.

At one public hearing, he recalled, "the mayor, a judge, and several attorneys spoke in opposition. Several doctors spoke, discussing their perceptions of elevated levels of cancer, respiratory problems, and [by an orthodontist] certain types of nasal deformities."

While the company used the standard assurance that its incinerator would burn at more than 99 percent efficiency (actually, 99.9999), Vestal calculated that at least forty-five pounds would still be released intact each year, or nearly a gram for each person in town. The EPA countered

145

that risk-assessment data from 1985 test burns showed that only .001 cases of cancer attributable to the incinerator could be expected over a seventy-year period—comparable to the risk of smoking one cigarette in a lifetime.

But the city's air was already tainted, the air sometimes a brown haze, according to residents interviewed by Bobbi Ridlehoover, the state's environmental scribe. They wanted a health survey to see if they'd already been affected—and they appeared willing to pay $250,000 of their own money rather than trust the state to conduct one.

Some of their cynicism probably came from the fact that the governor's wife, Hillary Rodham Clinton, worked for a law firm that represented ENSCO, and perhaps too from the charge that the very desk in Governor Bill Clinton's office had been loaned by a firm owned by the wife of Melvyn Bell, the incinerator company's chief executive.

The state was essentially siding with ENSCO in a dispute with the county, which had passed legislation barring the company from burning dioxin. The county's ordinance was eventually struck down.

Meanwhile, Bell was parlaying his position as toxic-burner-supreme into quite a base of power. He and some partners owned an amusement park and country club in Hot Springs, and he was also negotiating to buy the Savers Federal Savings and Loan Association of Little Rock.

At forty-eight he had accumulated what was reported to be $83 million worth of company stock, and he was known as a tough-talking, intimidating man who seemed to intensely dislike the press.

The governor's wife described Bell to me as "a friend, sure—he just gave $8 million to our engineering school at the university on Saturday.

"But he's not a personal friend, he's not somebody we socialize with, but he's somebody who's a friend—that's right."

Bell had not left behind such good cheer up in Minnesota, where he'd also run a site. It had been in Shakopee, where major environmental problems centered upon a huge fire that had sent waste barrels rocketing a hundred feet into the sky. "We had our taste of Melvyn Bell—that's putting it mildly," one Shakopee official, buildings inspector Leroy Houser, recalled bitterly.

Besides spills and other violations at the new site, one former employee claimed PCB warning labels had been removed from containers that were then sent to the Union County Sheltered Workshop for repair by handicapped people.

But Bell was treated quite cordially by Arkansas officials, and a former director of the state's environmental control agency had become a consultant to the company, which was now seeking permission to burn some of the dioxin wastes from Jacksonville.

Eventually ENSCO, citing market conditions, withdrew its application to increase the burning of PCBs—an apparent victory for the opponents.

The incinerator, however, already had taken an emotional toll. "Since I don't have home movies, let me just tell you the sights and sounds that I saw on my vacation last week out of El Dorado," said a local teacher, Sandy Gross, at a public hearing.

"I saw newborn babies being held down by nurses as they tried to inject them with their chemotherapy. I saw toddlers screaming, saying, 'Please no more spinal taps. Don't stick! No more back-sticks!' I saw children walking around with artificial limbs. I saw children in wheelchairs with no limbs. I saw children with radiation burns, and I saw my own daughter crying as they tried to inject her own chemotherapy in veins that are so bad now they can hardly even get it in.

"You see, the little trip I took last week was to Memphis to St. Jude's Children's Research Center. My daughter Christy, who was born here in El Dorado fourteen years ago, has cancer."

Her story was punctuated with sobs. "Christy has been through a living hell, and I'm not going to go through all the details with you, but I just pray to God that none of you have to watch your children or your grandchildren or anybody else go through what I've had to watch my daughter go through.

"The first day I took her to St. Jude's and they asked us where we were from and we said 'El Dorado,' two or three doctors said, 'You know, we have an awful lot of kids from El Dorado.'

"I asked the doctors, you know, what caused Christy's cancer, and they can't tell me, so I can't stand here and say that PCBs caused it, but I can stand here and tell you that if El Dorado doesn't do something to stop ENSCO, they're going to have to build a branch of St. Jude's here in El Dorado."

Through the hearing an EPA official who obviously saw it all as an overreaction wore a mostly amused smile. The cancer definitely did not jive with the .001 risk assessment.

At the same time, ENSCO's vice-president of engineering, apparently less amused than he was bored, doodled on a pad during the uproar.

Soon the people could see that he was doodling swastika signs.

147

This, then, is a glimpse of the Midwest's toxic kettle. Arkansas, Missouri, and the rest of the far Midwest not only stir their own seasoning into the main ingredients from Texas and Louisiana, but also into the thick sauce served up by western, windblown states.

Again, it is not as if America's breadbasket has been reduced to a mass of killer toxics, but neither is there much innocence left in the heartland—this massive coronary-respiratory system.

As for places such as Jacksonville, the people have been reduced to watching each new pregnancy with increasing trepidation.

The ghost of Joseph Shelton still hovers in their thoughts.

And though Patty Frase had not noticed it happening at Babyland, Mr. Park said that at Bayou Meto Cemetery, which is closer to the industry, something airborne is eating at the tombstones—and wearing off the names.

PART III

HAUGHTY HEARTS

17

There are skulls at the base of the Superwhip, cattle skulls, and one hangs at a ranch near Baton Rouge.

Another is nailed to the garage of Dave Haas Ewell, who used to hunt the swamps. "It's unreal what they done," he said, nodding to the refineries and chemical plants that crowd the Mississippi's banks. "And ya know what? It's our children who'll pay for it. They will."

His family's herd was poisoned by chlorinated hydrocarbons, solvents, and metals which had deluged the area years ago, scalding the cypresses, turning alligators belly-up. They came from a chemical plant next to his family's plantation, and I remembered that upon my first visit there, in 1979, shiny globules of mercury oozed from the indentations my foot made in the muck near a bayou.

The trees were lifeless, devoid of even mosquitoes, the sand and bark giving off an ethereal odor. Nearby, whole tank-cars had been buried and chemical residues had been spread just below the surface and repeatedly plowed so bacteria and sunlight disassembled them—a disposal method known, incongruously, as "landfarming."

When the sun was hot, a thick, black sludge surfaced down by the bayou, sending an oil slick toward the Mississippi, which supplies drinking water to the city of New Orleans. There the environment and cancer rates are such that Dr. Velma L. Campbell, a physician at Ochsner Clinic, describes them as "a massive human experiment conducted without the consent of the experimental subjects."

In the graveyards here, where tombs are raised above the soil because of high groundwater levels, there are other skulls. Epidemiologists have not yet proven any of these the result of what man does to the

151

atmosphere, but certain evidence is falling into place, and these bones, of course, are human.

The situation in Louisiana is summarized best by the name of the contaminated area behind the Ewells's property. It is known as Devil's Swamp. Soon, unless drastic remedial measures are undertaken, it is a name that will aptly apply to much of the state, as well as some of the other coastland at the base of the Gulf Superwhip—in Mississippi, Alabama, and most especially Texas.

This is one of three major foundations of air that feed the Midwest and then move widely east. The other two are the Canadian Cool and the westerlies.

Hence, a little of the pollution we see here can be expected to get up as far as Minnesota and Wisconsin, perhaps find its way over the Great Lakes, over Pennsylvania, or end up at sea, the molecules of carbon streaming in infinitely small patches above the flickering lights of New York.

But in addition to telling us what is circulating in the masses of air, Louisiana, more than anywhere else, gives us a striking glimpse of the relationship between chronic disease and proximity to the smokestacks. The heavy exposures Louisianians absorb give a quick preview of what lower-level exposures elsewhere may one day cause. In this sense, the residents are coal-mine canaries.

In May of 1985 the EPA's Office of Air and its Radiation Office of Policy, Planning, and Evaluation released a study entitled, "The Air Toxics Problem in the United States: An Analysis of Cancer Risks for Selected Pollutants." In the introduction its authors made it plain that their work was inspired by brief congressional hearings in 1983 that examined Section 112 of the Clean Air Act, which requires EPA to protect the public from exposure to hazardous pollutants—a requirement that EPA, many felt, had blatantly ignored. Since passage of the act, EPA had listed only seven substances as hazardous air pollutants and established emission standards for but four of them.

As if to excuse the agency's behavior, the report, a mere hundred pages in length, indicated that there was not yet evidence of any acute crisis. It seemed geared to buttress a statement made at the congressional hearing by the administrator, William D. Ruckelshaus, who said that "a very preliminary evaluation" of air toxics indicated that the annual excess cancer cases from them were probably "a few hundred rather than thousands."

But at the same time, the report's very commissioning signaled federal concern about the possibility, however unlikely the administration hoped it was, that, indeed, a new, very major category of environmental distress might have been allowed to slip through the cracks and could one day rise up to smite the agency with a loud, embarrassing, and monumental controversy much as the issue of hazardous land dumps had done in the wake of the Love Canal crisis.

In and of itself, the study received very little publicity, its lasting fame limited to the inner circles of federal and state air-pollution regulators. Some believed it was a first step in the right direction, a significant beginning in figuring out if air toxics are a real problem, while others, like Congressman Henry A. Waxman of California, felt it was "riddled with errors and omissions so sweeping that it has been a giant step backward."

Whatever the final verdict, there was unquestionably some valuable data in the study. For example, it began to place into perspective what airborne compounds were of most immediate concern to us. (It gave a very high rank to our familiar friend benzene and to the metal chromium, a carcinogen used in a wide variety of industries including those that make magnetic tapes, paint pigments, cement, paper, and stainless steel. And it said that a lion's share of the compounds were manufactured between New Orleans and Corpus Christi, Texas.)

But the study's greatest value was in leading its readers to realize how little was known. Its own conclusions, admitted the report, were fraught with "major data gaps and assumptions," a degree of uncertainty, in fact, "that we cannot begin to quantify."

Only fifteen to forty-five chemical substances were analyzed—or less than half the carcinogens identified by the agency's Carcinogen Assessment Group and a mere .075 percent of the total compounds that are regularly being produced and therefore might be getting into the air.

While a good number of the nation's most poisonous and most widely used chemicals were surveyed, dioxin and furans were not among them. With what data it *did* compile, however, the authors indicated that, nationally, thirteen hundred to seventeen hundred cases of cancer might be expected each year from air toxics.

Near places like Devil's Swamp, and down the corridor from Baton Rouge to New Orleans, which is composed of about a hundred factories and refineries, it seemed like all of those cases must have been occurring in Louisiana. While, in general cancer occurrence, several Northeastern states still had higher rates for the 1970s (reflecting, perhaps, a longer exposure to industry), Louisiana lost 7,636 people to cancer in 1983 alone, and the decade before, it led the nation in lung cancer deaths with a rate of about eighty per 100,000 people—about 10 percent more than runners-up Maryland, Delaware, and Mississippi.

The biggest splotches of black on the nation's maps of lung cancer rates had moved far south, crawling like sludge nearly halfway up the Superwhip's stem and allowing the lower Mississippi to borrow from New Jersey the nickname "Cancer Alley."

Actually, because it follows the quarter moon of Gulf coastland from western Florida to east Texas, the blackness on cancer maps is more accurately described as a huge crescent. And black it is. In one study of white males in three thousand American counties, thirty-eight of Louisiana's sixty-four counties (or parishes) were in the upper 10 percent. During some surveys, cancer of the pancreas and bladder also seemed quite prevalent. Noted a special report to the governor in 1984, "Nearly one cancer death per hour in Louisiana is reason to be gravely concerned."

The reason: There were only 4.4 million people in the state.

In St. Bernard, the parish with the highest rate of all types of cancer mortality for white males in Louisiana during the last decade, the rate was more than 35 percent higher than both Midland, Michigan, and the state of Missouri.

The statistics can be made to call forth with an even shriller voice. In one recent period, for example, figures showed that 30 percent of state residents could expect to develop cancer over a lifetime, including 33.1 percent of those living in New Orleans.

While white females in Louisiana could take some comfort in an overall rate lower than the national occurrence, the incidence for cancer of the mouth in white males, to cite another specific category, was 169 percent higher than the U.S. rate during a different period of review.

Black males in the New Orleans area led the *world* in new cases of lung cancer, while the lifetime risk of contracting cancer for white males there stood a jolting 40.1 percent.

The statisticians and EPA risk-assessment specialists said little about the real ramifications of these cancer figures. On paper the numbers were officially "suggestive" or "significant"—but nothing that provokes the type of urgency which comes about when one visits the homes of people like Gay Hanks, who lives in the tiny town of Kaplan in the south-center of the state, and whose daughter, Angela, presented a much more relevant set of feelings than any which may be evoked by rat fetuses, cultures of salmonella, or autopsies on rhesus monkeys.

Kaplan is a Cajun town, and the Cajuns, who had come from Nova Scotia in the eighteenth century, had been quite put upon by the oil men who surged into southern Louisiana during the 1930s and turned a culture which had revolved modestly around catching crawfish and trapping muskrats into an industrial morass of quick money, galloping cancer, and sliced-up swamps.

In a chilling parallel to the Sandra Gross testimony in Arkansas, Mrs. Hanks too had found herself taking Angela to St. Jude's in Memphis, and she too had heard a doctor say that there were too many cases from her part of the country—enough so that they car pooled the ten-hour drive up to the cancer ward.

But there were also some major differences between the two cases. For one, Angela Hanks had been there—and died there—twenty years before the Gross youngster.

The images will always hurt Mrs. Hanks. "Someone once said losing a child is like going the rest of your life with one shoe on and one off. And it's true. It's amazing the profound effect it has. She was our thermostat. She was vivacious. How she felt we all felt."

She reached for some pictures despite the threat of tears. "Here's Angela."

She showed me a photograph of her daughter on a bed with stuffed animals, at the age of five. That was about a year before her death. And there was also one taken after her hair had grown back in the aftermath of chemotherapy.

"She would lay on a pillow at the front door and watch the other children play. She didn't have the energy and she was pale, but she started school anyway. Her blonde hair was everyone's pride and joy. She was like a little blond angel. And when she lost it, it was even more traumatic. It devastated her. She was so ashamed.

"When they did the bone marrow tests you could hear her scream across the hospital," Mrs. Hanks continued. "And the fear: you put a four-year-old child in a room and put a needle into her bone and you're going to have a frightened child. The memory of those screams . . . I can still hear them."

155

But more often, Angela was a profile in courage. She did well at school. She would have been a very bright child.

"But she didn't have aspirations. She knew she wasn't going to grow up. One day at the hospital she looked out the window at the clouds and said, 'I'm gonna be a saint.' "

As Angela lay dying, her mother left the room because there was just too much pain to see.

We went to the cemetery up the road so Mrs. Hanks could show me the grave she visits each day. The clouds above were heavy and still, and Mrs. Hanks, who is now an ardent environmentalist, took control of her emotions and silently prayed over a raised white-stone tomb decorated by the statue of a little angel.

It was not the type of sight that an "objective" and unemotional scientist or bureaucrat would want to see. To such people Angela and the other white females who died of leukemia in Vermilion Parish during the 1960s looked in an atlas of cancer incidence only like this: "10—5.6," meaning 10 cases at a rate of 5.6 per 100,000—nothing very special.

(In a column above, however, for the following decade, was the number 856*. It was the number of cases for Louisiana as a whole, and the asterisk meant there were too many cases—statistical significance again.)

It was in the 1950s and 1960s that medical researchers began tracking blips that turned out to be sharp and steady upward lines on the cancer graphs. Since cancer is believed to take up to two decades to blossom in the body, that takes us back to the 1940s, when, coincidentally, the petrochemical industry was taking deep root along the lower Mississippi, drawn there by cheap feedstock gas from the oil fields, easy water transport, and lucrative tax incentives.

The government of Louisiana serviced the chemical makers as no other government did—especially those that were offshoots of the revered petroleum industry. A manufacturing company could obtain up to ten years of exemptions from property tax in the state, and as the older industrial states in the North began controlling the wanton pollution there, corporations looked to the South for relief from such costly regulations.

Soon companies making pesticides, chlorine, rubber, antifreeze, detergents, plastics, and just about everything else sprouted along the river and into the swamps. The tankers rolled from New Orleans to Lake Charles. Virtually every major American chemical corporation in the country has an outpost along the river, and some foreign ones do too. By 1986 Louisiana provided a third of the nation's nitrogen fertilizer, 23 percent of the ethylene, and was ranked third nationally in chemical production, behind New Jersey and Texas.

There are many other kinds of industries that contribute to cancer, and by one estimate in the EPA report, only 4 percent of the cancer incidents caused by airborne toxics can be laid directly onto the lap of chemical manufacturers, while a nearly equal number come from metals processors or from a combination of emissions from the rubber and petroleum industries.

Due to the products of incomplete combustion, road vehicles by far account for the most cases of toxicant-induced cancer.

But for the most part the chemical firms send out the most troubling types of molecules, and in Baton Rouge, according to company data, an Exxon Chemical plant was leaking 560,000 pounds of benzene yearly, while just south of there, according to a survey by the Sierra Club, eighteen plants in and around St. Gabriel and Geismar dumped about 400 billion pounds of toxic chemicals into the air during the first nine months of 1986.

The implications of such extensive chemical production are far beyond any cursory review. In 1982 Louisiana accounted for about $11 billion of the $170 billion in chemical shipments nationwide. When the outputs of both Texas and Louisiana are taken together for petrochemicals (those products of oil and gas hydrocarbons, as opposed to those that are based upon metals, salt brine, and substances such as sulfur and phosporous), the two states represent a towering *half* of the nation's production—actually, 60 percent.

Louisiana, in area, is not a large state, ranking thirty-first in size; and so there is not much room for either the large refineries or the residues they spew. Nor is there much room for the wastes that Louisiana *imports* from the Midwest and other regions for disposal. In 1983 the state accepted 305.6 million pounds of out-of-state wastes.

The legacy of the oil industry in Louisiana also includes fourteen to twenty thousand oil-field pits, which may contain not only oil and brine, but acids, barium, and lead as well. In fact, more than four hundred other compounds can be found in the drilling muds used to extract petroleum, including cobalt, chromium, radioactive potassium, and caustic soda, the active ingredient in Drano.

At one point, Louisiana could lay claim to three of the nation's four largest hazardous-waste disposal sites.

By another calculation that obviously categorized more materials as "hazardous" than the national estimates did, the state, one recent year, generated sixteen thousand pounds of such wastes for each of its citizens.

Given the evaporation from these pits, the volatization from ponds of waste liquids and from storage tanks, the miasma rising from places such as Devil's Swamp—the steaming, hissing valves at petrochemical plants all the way down the Mississippi, and the plumes of refinery smoke that form their own cumulus strata across the horizon—the picture in southern Louisiana becomes one not of a low, lingering smog but more that of a toxic tornado.

No state has more pressing, more devastating problems. Even hardened industrial consultants and government officials come back shaking their heads. In community after community, house to house, are stories each of which is the equal of those in other states that have received the kind of national attention Louisiana has not.

Louisianians rarely make it into the limelight, and the EPA does not seem to like venturing into its swamps. The agency is perhaps fearful that if it were to thoroughly investigate and expose each toxic threat there, the state as a whole might come to be viewed as the very worst environmental disaster in the nation's history.

At the same time, however, the state had cut its contribution to the Department of Environmental Quality (DEQ) by 80 percent of what it once was, and it was looking to snip off 30 percent of what was left of that. (Fortunately, most of the department's funding is from the federal government or is raised by fines and fees.)

Per capita, Louisiana spends a sixth of what many other states do on the environment—a scrawny two dollars or less a head.

Some of this is understandable in that the state's income has shrunk as oil prices have plummeted. Unemployment rates of 12 percent have become commonplace.

But looked at another way, less than $1 million has been spent annually by the state on cancer research, a fraction of what a major cancer clinic might spend.

Also, a bill that called not for increased spending but rather for larger fines against consistent environmental offenders—which would have brought in more revenue—was once killed unceremoniously.

The best symbol for the state of this state is the fact that chlorinated hydrocarbons, specifically endrin, had been suspected as having led to the extirpation of the brown pelican. This was especially embarrassing because the state seal is that of a mother pelican feeding her brood. License plates in Louisiana carry the slogan "Sportsman's Paradise," and the state's very nickname is "The Pelican State."

Another symbol of the environmental crisis are the signposts planted in Capitol Lake, the big pond next to the governor's mansion.

The signs say, "No Fishing—Cam Cau Ca."

It is so the Vietnamese immigrants would also know to stay away, for PCBs are present at eighteen hundred times the federal water standards.

Not far away is the Bayou Montesano. It trends through the pipeline jungle of Baton Rouge, turning from blue to emerald to olive green. In this stretch of industry one could find such names as Ethyl Corporation, Stauffer, Copolymer, and Allied. A bit down the river was Georgia-Pacific. The state's attorney general and the DEQ, which was under a bold, rebellious administrator named Patricia Norton, claimed that this plant had released more than 55,723 pounds of vinyl chloride; they also intervened in a federal case in which it was alleged that Ethyl Corporation released 183,218 pounds from 1977 to 1983.

Throughout the Baton Rouge area the sky is red with the type of pollution that is 30 percent higher on weekends—when the few DEQ inspectors are off work—than during the rest of the week.

It is the type of turf where "mystery clouds" that no one can identify appear as if from nowhere, and where white fallout and gushing flares can be seen.

I obtained one preliminary laboratory report which showed that a catfish taken from a fish farm near the hamlet of Pigeon was contaminated with stunning levels—more than ninety-one hundred parts per trillion—of TCDD.

The stretch down to New Orleans presents a cornucopia of sheer environmental travesty. There are other dead swamps, and near one, in St. Gabriel, I was met on the narrow entrance road by a large Brahma bull which shook its horns, challenging my car.

Eventually the animal wandered into a forest of stark cypresses which had been killed by oil-field brine.

Not far from there were mountains of radioactive gypsum, contain-

159

ing low-level radium and uranium and surely sending dust into this great toxic engine.

Near one refinery in the awkwardly named town of Good Hope, homes had been boarded and abandoned, taken over by a massive refinery that wanted the land and was tired of citizen's complaints about its smoke.

Eventually the wooden homes were hauled out on flatbeds, and except for a few barrooms and a Presbyterian church, the place, like Times Beach, has basically ceased to exist.

"There were fires, air pollution, catalysts erupting from the catcracker onto our homes, we had very many odor problems," said a mournful Charles Robicheaux, who had been born in friendly, tightly knit Good Hope but had to flee with his eight children.

"At night the flare would rumble and actually vibrate our beds."

The great leeway granted to industry is a barometer not only of private economic greed but also of corruption in both politics and bureaucracy. The most obvious target of such allegations has been Governor Edwin Edwards, whose cordial, comical rhetoric and flamboyant gambling sprees have garnered him both amused respect at home and publicity nationwide.

The governor's closeness with industry is a volume unto itself. While he was drawing $100,000 a year in oil and gas royalties from Exxon Corporation, Superior Oil, and other firms, he was telling citizens and reporters that environmental mishaps are "trade-offs" that are necessary for the sake of economic health.

A lawyer, in the hiatus before his third term he had legally represented a number of oil interests. And it was oil money that fueled the most expensive gubernatorial election in U.S. history. While the state was heading quickly to the poverty level—its oil and gas nearly depleted, and the profits from its oil rush in the hands of companies (many based out-of-state) which are now pulling up their stakes—Edwards could nonetheless afford to spend $15 million campaigning.

According to the *Wall Street Journal,* William Huls, for a while his secretary of natural resources, abruptly dismissed an accounting firm which had been doing an audit of Texaco's royalty payments to the state. The accountants said afterward that the oil giant owed financially strapped Louisiana at least $100 million.

Also, Edwards had once appointed Raymond Sutton, a former salesman for a waste-hauling firm, as state conservation officer.

Just before my last visit to Louisiana, Edwards was acquitted in a federal trial of fraud charges. More recently, on October 16, 1986, the governor was called before a grand jury investigating allegations that state pardons had been sold.

The conflicts of interest came by the barrel. One oil-field regulator was actually advertised on business cards as a consultant to a chemical sales business! Another had an Amway sales group. One of his distributors had sold forty thousand dollars' worth of the detergent products to drilling contractors and others in the oil trade.

On the Louisiana State Mineral Board, which dispenses oil leases, had sat the owners of several firms which *serviced* the oil industry.

Lax regulation and unbridled production are a recipe for both immediate disasters and chronic harm. In Taft, a town within five miles of Good Hope and twenty miles northwest of New Orleans, a chemical explosion in 1982 drove seventeen thousand people from their homes, an incident standing, in retrospect, as an ironic harbinger to the tragedy in Bhopal, for the plant on the Mississippi was also owned by the Union Carbide Corporation.

Other big chemical names—Du Pont and Olin—had also suffered mishaps; the trade unionists had accused the German-owned BASF Wyandotte plant in Geismar—which handled phosgene, isocyanates, and other compounds similar to those which had been produced in Bhopal—of "numerous chemical releases" and "operator errors."

Such accusations were difficult to fully evaluate, coming as they did in the heat of a labor dispute. But other facts made themselves more obvious. On September 28, 1982—back up near Baton Rouge—a derailment of forty-three tank cars containing four million pounds of toxic materials caused the evacuation of twenty-five hundred residents from Livingston for two full weeks. A series of explosions flung the cars around the middle of the town, noted the Baton Rouge *Morning Advocate,* "like toys discarded by an angry child."

Elsewhere there were escapes of tear gas, herbicides, and green, glowing clouds. They came from valves, pipe welds, storage tanks, and the tanker trucks that rumble into Baton Rouge at the rate of nearly one a minute.

When I asked Patricia Norton, secretary of the DEQ, if a Bhopal-type incident could happen there, she said, "Yes. I don't think anyone could tell you it couldn't. We've come dangerously close."

161

But more worrisome are the smaller, steadier day-to-day exposures. In New Orleans I heard accounts of the air breaking down nylon hose and burning holes into synthetic sweaters.

In one area the contamination with asbestos was so bad that a doctor was quoted as saying, "When I drive down Fourth Street I hold my breath."

At a picnic with local environmentalists near the Audubon Zoological Gardens I also heard stories of rolling industrial smoke causing highway motorists to stop their cars in utter confusion; and tales of fumes coming off the polluted water in shower stalls.

Among those at the picnic was a sullen surgeon whose wife was in the hospital dying of cancer. Also present was Dr. Victor Alexander, a Harvard graduate who had served as a senior medical officer at the federal Occupational Safety and Health Administration (OSHA) in Washington before moving into the living laboratory of Louisiana, where he loudly voiced concern over toxic air pollutants and where—rightly or wrongly, in a bizarre, incomprehensible turn of events that he claimed was a mistake or a frame-up—the physician had been arrested and convicted of a local bank robbery.

Louisiana!

At each turn there was something of the renegade in this state. Way up near Ruston is a creek where oil-field wastes had been dumped and caught afire—the blaze racing down the waterway and destroying a bridge.

Sportsman's Paradise!

After the New Orleans–Baton Rouge corridor the most troubled part is to the far west in the Lake Charles area, where smelly smokestacks send small clouds up the Superwhip, some days straight toward Arkansas, Missouri, and Iowa, other days bending more easterly toward Mississippi, Kentucky, and Illinois.

The situation is perhaps best understood in knowing that a few local kids in these parts have been found to have significant levels of benzene in their blood. According to local activist Shirley Goldsmith, the area is also incurring the standard plagues of cancer.

But in the opinion of many authorities, the state's extraordinary cancer incidence is related much more to individual life-styles than to carcinogens in the ambient environment. Frequently, the finger is pointed at the Cajuns in southern Louisiana, who, it is claimed, tend to smoke unfiltered cigarettes or roll their own strong ones, drink too hard, and eat improperly.

In some parishes, according to the Lung Association of Louisiana, 40 percent of the adults smoke tobacco versus about 30 percent in the rest of the country. Overall, says Dr. Pelayo Correa, head of the Louisiana Cancer and Lung Trust Fund, inhabitants of the state's southern section smoke 5 percent more than those to the north, where the smoking is about in line with national averages.

In addition, noted the Louisiana Chemical Association (LCA), high rates of cancer actually had begun during the 1930s—before the chemical industry's boom times.

To no one's surprise, the LCA is one of the most vocal proponents of the "life-style" theories.

No one can doubt that smoking is the overwhelming cause of lung cancer, a fantastic and unprecedented spur to disease; and two prominent English researchers, known in the literature as Doll and Peto, believed pollution causes only 2 percent of all cancer deaths.

Others feel, however, that the impact of environmental contamination—especially airborne contamination—is much higher than that. What's in the air alone may cause 20 percent of lung cancers, I've heard it said, which, in Louisiana, would mean hundreds of deaths a year from that form of cancer alone.

There is the growing feeling that a very important factor is synergism. The combination of two or more cancer-causing exposures (such as smoking *and* chemical inhalation) may multiply the effects of each other. Such has proven to be the case with asbestos: Smokers are twice as prone than nonsmokers to the fiber's deleterious effects.

At the same time, a study by the National Cancer Institute has shown nasal, sinus, lung, and testicular cancers are higher in counties that have some form of petroleum industry.

Another Harvard-spun researcher who had migrated to the South and caused some controversy is Dr. Marise Gottlieb, a well-known cancer epidemiologist teaching at Tulane University's School of Medicine. Dr. Gottlieb studied people living within a mile of large chemical plants, and she found they were 4.5 times as prone to lung cancer as those further away.

She also had found what appeared to be a correlation between lung cancer and the shipbuilding industry. But as she drew ever nearer to specific causal links in environmental settings, her funding dried up. "I was told very bluntly they [the state] didn't want the studies on cancer," she says. "I did what I could. I came down, tried it out, wanted to contribute, and couldn't."

163

At the same time cancer researchers were being frustrated in proving a final link, preliminary data gathered quietly by state and EPA technicians already indicated a very serious problem with carcinogens and mutagens in the ambient air. "You know, it's unbelievable what they found down there," one EPA scientist confided to a reporter. "It's just so kinky."

Certain samples taken in and around Baton Rouge, including in front of the capitol, showed the existence of a breathtaking array of volatile organic compounds—in one case more than eighty of them. The maximum reading for benzene outside the state capitol was more than the maximum found *inside* homes during initial tests at the worst section of Love Canal. (Those homes, of course, were soon evacuated.)

While other values were lower than those near the famous dump, they sometimes approached levels at Love Canal for compounds such as the solvent chloroform; many samples were in the part-per-billion range. And more than thirty halogenated hydrocarbons were identified in the ambient air of Plaquemine, Geismar, and Baton Rouge, at vapor levels that at times totaled more than ten thousand nanograms (or ten micrograms) per cubic meter.

In Lake Charles, at a City Services refinery, toluene and ethyl benzene were found in the air.

One EPA official, Allyn M. Davis of the agency's Dallas office, says that there are probably a thousand chemicals lofting into the air and combining to form a total of perhaps *ten thousand* compounds. During some of the sampling, fumes were so strong technicians had trouble staying long enough to install their monitoring devices. They also were nervous about the dead cattle they spotted close by.

Along the petrochemical shore in Baton Rouge the street signs and store lights are bleeding rust—paint peeling here too, the ozone almost palpable.

In the air hangs a smell like that of oil laced with solvents, followed by that of rotten potatoes.

Further up were Allied, Kaiser Aluminum, Exxon Plastics, and a huge, sprawling tank farm. Most significant, however, was a site called Rollins Environmental Services, where, by a 1980 count, the most toxic wastes from sixteen states and as far away as Puerto Rico had been dumped into pits, landfarmed, and incinerated, the disposal amounting to 386 million pounds that year alone.

It was also where the cattle skulls were. Some of them came from a herd that had died back in the 1970s, poisoned while they were wintering near a bayou. The waterway runs through Devil's Swamp, which had been inundated with toxic sludges from a second disposal operation to the north. As for Rollins, even workers at another chemical plant complained about its odors—and filed suit in an effort to stop them.

Right in the very middle of the chemical deluge was the Ewell Angus ranch, which still carried the flavor of an old-time plantation, with hummingbirds flitting in the bushes and with the lady of the house, Catherine, cutting a dainty, elegant figure.

Her family told me of the chemical dust which had fallen on the vegetation, of the windows that had to remain closed for years, and of their burning eyes. "You couldn't sit here with the doors open," she said. "I thought I'd have to wear gas masks. [They] were landfarming and the odors were bad. I thought I was going to have a nervous breakdown."

According to William A. Fontenot, an environmental investigator with the attorney general's office, when soil was removed at one dump, the chemicals were found to be reacting with each other—sending off fumes as they literally boiled under the ground.

Back in the swamp, fish with swollen bellies regurgitated white pellets when they were stepped upon by sportsmen, and the crawfish had black stuff in their heads.

A chemical smell came out of fish frying in the pan. Others told me of the peas that were plowed under because the blossoms fell off, of the wafts of white, black, brown, and red smoke, of the terrible tumors on the swampland fish.

There were all kinds of sources in the area.

"Oh, my God, it was so bad in the morning when the sun was coming over the trees you could see rays of pollution," said Annie Bowdry, coordinator across the highway for a community center in the tiny poverty-stricken, sharecroppers' hamlet known as Alsen—unlike the plantation, a community of blacks in trailers and small homes.

On February 6, 1986, an emission from the direction of Rollins—one that looked to a witness like a fog cover bringing rain—caused schoolchildren and employees in the vicinity to gag and turn nauseous.

During other releases the citizens claimed to have staggered around blinded and choking. One of the residents suing the company for its fumes, Emma Johnson, a woman in her mid-sixties who'd had respiratory problems, was taken to the hospital after an odor episode and died not long afterward.

Her husband, Leon, who had been there since 1938 and remembered when the Rollins property had been farmland, was a slow-talking man nearing eighty who had a portrait of him and Emma sitting side by side in two chairs. There was also a large Bible near the couch—a companion in times of loneliness.

"It got her real bad," said Leon. "She couldn't stand that odor. She could smell it when I couldn't."

Back near the swamp itself, at a dead-end sign pocked by rifle shot, was the home of Elree Pate—"Brother Pate" to his cronies. His home was often in line with the Rollins emissions, and he described a blue haze and pinkish-orange smoke.

The house he once had considered his dream house was now for sale—with no takers yet—and there was a red skull on a post in his yard, along with the prized trees he had planted as saplings but now planned to abandon because of the fumes.

"I've only got a sixth-grade education, and to me, this was my goal, where I would raise a pig or a few chickens, have a vegetable garden, to me it was a big accomplishment. But I can't enjoy it. I got some friends who won't visit anymore."

He pointed to the "steam" rolling out of Rollins. "They should have had this thing in a desert, some remote place where nothing grows or lives anyway. We've woke up in the middle of the night, my mother coughing and choking. We run into her room and when we opened her bedroom door, man, we started gagging, it was that bad, it had come through the air conditioner—like burnt leather, burnt flesh, or insecticides.

"We got in the truck and drove around until it cleared out."

He showed me a stack of complaints to environmental officials.

One of them, DEQ Secretary Pat Norton, had come out to Pate's place and the odors nearly caused her to collapse.

Norton was a divorcée in her early thirties, a backpacker who prided herself with being in tune with nature, keeping elderberry branches in her car instead of artificial fumigants to keep the fleas away.

A lawyer who had once belonged to the Sierra Club and the Wildlife Federation, Norton had worked for the attorney general's office on cases involving Devil's Swamp before Governor Edwards—in a concession to the state's increasingly concerned and protesting citizenry—appointed her to head the DEQ.

He could not have known how tough she would be. She proved her mettle in August of 1985, when, after another visit out toward Devil's Swamp, she shut the Rollins plant down.

The visit was on August 5, after the DEQ heard complaints of yet another attack by Rollins's fumes. In its six years of history there, said Ms. Norton, Rollins had "one continual problem after another," and according to company logbooks I saw, the month leading up to August 5 was one in which every other day seemed to bring another problem to the incinerator, as if the operation was one of patch jobs that were barely able to keep it running.

There were line pluggings and backfiring and hose leaks and fallout from the stack. On July 15, 1985, a log entry read that "fallout and smoke is terrible coming from the stack" and that a line to the kiln had to be shut "so we can breathe."

This was startling information considering the high potency of what was being incinerated and the fact that any incomplete combustion of chlorinated (or otherwise halogenated) materials meant furans and dioxins might form.

A downward fluctuation in temperature might be all that was necessary to loose them into the air, and if the emissions from the stack were anything but steam, that raised the possibility of incomplete combustion and toxic particulates.

On August 5 a very black plume of smoke was spotted coming from Rollins, and fumes were such that an annual summer revival at Mount Bethel Baptist Church had to be cut short, the pastor having trouble getting the words out of his throat ("I couldn't shout like I wanted"), having trouble reading his Bible. When state investigators burst onto the waste site they found the incinerator control room out of control.

One employee seemed to be dizzy, faint, his limbs jerking; and according to Norton, the "operator couldn't explain what was wrong. Since then we've theorized that [they] misconnected a valve or left a valve open, causing an imbalance in the feed."

Also, the monitoring equipment was not working. "They seemed to have an attitude of, 'We've got to keep this running no matter what. It doesn't matter if a couple little pieces of equipment break, it doesn't matter if we're in violation of the law.' "

When I asked if she felt the Rollins site was affecting the Alsen residents, Ms. Norton sounded angry. "I know one of them who died as a result of it," she said, referring to Emma Johnson. "She was a witness in the case I prosecuted four years ago, and when I had her on the stand she testified that her doctor told her if she did not get out of the fumes or the fumes didn't stop, that she was going to be dead.

"And within six months, she died."

Secretary Norton's actions against Rollins evoked a highly unusual response. The company sued the public official personally in what she interpreted as an effort to harass her and make her back down. The governor himself told her to retreat, and after Ms. Norton was removed from the case in a court decision, Rollins hired Bill Broadhurst as a legal counsel.

Broadhurst happened to be Governor Edwards's close buddy and former law partner.

Rollins had also hired Dan Burt, a short, caustic attorney from Washington whose notoriety had been won in a lost case: the Westmoreland trial involving CBS. He was a self-described legal "hit man," and Burt now had turned the full force of his venom on Ms. Norton, to the point of offering information about her sex life.

He was even unnerving to lawyers at the attorney general's office, not to mention the small private practitioners who were representing the victims. One of them, Stephen Irving, seemed quite battle-weary when I met him in his small, cramped office in Baton Rouge. He said he too had been personally attacked by the Rollins legal team, which he accused of "gutter tactics." He was especially angered because, he said, Rollins had hired Lee Wesley (former director of a local antipoverty agency who had been found guilty of conspiracy and misapplying federal funds) to go around and try to settle with the defendants behind Irving's back.

One of those said to have been offered money—Mary McCastle, who lived near Emma Johnson and herself suffered from emphysema, though she did not smoke—claimed to me that right after she refused the offer from Wesley she was laid off her job at a city food program for the low-of-income.

Louisiana! It had always been wild and wooly, a state that one official there described as "the northernmost banana republic." As for Rollins, the attorney general, William Guste, described it as "a terrible corporate citizen," a "three" on a scale of "zero to ten."

"Let's say that Rollins snickered at Louisiana environmental regulations," he said, rethinking the rating he had given it. "You'd better make that a 'two.' "

Controversy seemed to attach itself to Rollins like hydrocarbons latching onto particulates. According to a decision by a federal appeals court in 1982, Rollins once arranged with another company to secretly dispose of highly hazardous wastes by mixing it with crude oil for sale through a pipeline. It caused a major explosion after reaching an Ashland Oil Company refinery in Kentucky!

Because Rollins has been described as the nation's leading operator of incinerators, its controversies are of more than passing interest. For instance, there were the toxic wastes that fell from a kiln near Deer Park, Texas, and smoldered on the ground; and the waste site in Logan Township, New Jersey, that exploded into a toxic inferno years before.

Rollins gave bonuses according to production levels, so it was burn and burn some more: Louisiana officials got an anonymous tip at one point that emergency equipment had been tampered with so that the incinerator would stay on even under adverse conditions.

Yet for that August 5 incident which stopped church services, Rollins was fined a paltry fifteen thousand dollars and was considering performing a public service in lieu of the fine.

The public service would be to install an air conditioning system at the Alsen community center.

A few months after I accompanied an NBC news crew to Louisiana and Norton, in a series of poignant segments, spoke out candidly on television, telling NBC what she'd told me, Governor Edwards fired her.

Meanwhile, said Attorney General Guste, the Rollins verdict stood as "a disgrace, an affront to the people of the town of Alsen and all Louisianians."

But more than a simple affront were the varied and peculiar hydrocarbons that rose from many parts of the state to join Texas in spreading poisons up the Superwhip toward unsuspecting parts of America.

18

"I will never forget it because it was the darkest day of my life. The sun stopped shining and it got dark like night and it got cold. It was around 9:12 in the morning. We heard that terrible explosion, just one after another. People were screamin' and runnin'. Hot iron was flying around in the air. I seen several pieces that hit the ground.

"We went out in the yard and saw an orange-colored smoke. And you'd find a leg there or an arm here. I lost a lot of friends. Some of them were close friends. The woman who lived in a house that was here started out her side door and a piece of iron just about the size of your two hands flew and decapitated her.

"Her head was one way and her body the other. And her husband, he was killed at the plant."

The date April 16, 1947, is often remembered by Rachel Chargois. It is the day the *Grand Camp,* a French freighter docked near a Monsanto facility and filled with ammonium nitrate, exploded with such power that it blew out the grinding, buzzing factory that is immediately behind her home. It also threw a 10,640-pound anchor a half mile across town, near the site of the present-day Texas City Inn. A second and third ship blew in the wild chain reaction, an accident in the class of Bhopal. Five hundred seventy-six were killed and five thousand injured.

Long ago though it was, the catastrophe remains indelibly etched onto the memories of those whose homes abut the enormous refineries and chemical factories on Texas's Gulf Coast. These plants, the nation's largest, are much too close for comfort, and their sounds—the sirens,

171

the horns, the relief-valves popping like gunshot—are heard like a neighbor's loud stereo set.

In the farther background are shrimp boats and smoke plumes and stalls of mudfish.

Located on Galveston Bay, less than an hour's drive southeast from Houston, Texas City is one of many coastal towns that are not simply heavy with industry but totally possessed by it. The factories are living, fire-breathing serpents of steel. Horses trot right up to the plant's fence-line, sniffing alcohol instead of hay, stirring at the sound of a tanker; and at night residents watch shadows flicker about their bedrooms, cast there by the raging smokestack flares.

Besides Monsanto, such corporations as Union Carbide and Amoco also reside in the area, waking the residents with an occasional roar. The soot falls in hues of gray and green, smelling here more like sauerkraut.

The industrial toil causes complaints too of molten spills and dead songbirds and in one case of a bride vomiting on her wedding day.

At Fannie's Diner, old industrial drums are padded on top and used as stools.

While the danger of an explosion is a prime concern, it is beginning to take a backseat to worries over the insidious toxic fumes. From Corpus Christi through Houston, northeast to what is called the Golden Triangle (Port Arthur, Orange, and Beaumont), is what amounts to America's main industrial exhaust system.

Ferns turn white and small mammals lay dead near a leaking pipeline.

Outside of Houston, bands of reddish-brown smog are actually visible some days for sixty miles: a caravan moving styrene and nitrates and all sorts of volatile organics due north.

They follow the trends of the high-pressure system, moving air clockwise, that often hovers over the Atlantic and Gulf.

So big are the industrial complexes in Texas that they have been known to cause a phenomenon called the "heat-island" effect, creating clouds of warm air that rise like the air in a low-pressure center. In other words, the tangled structures built by man are cranking up their own miniature, counterclockwise weather systems.

They also appear to be causing new cancer patterns. This part of the crescent is getting very dark too, with rates of lung cancer that rival and in a few specific instances surpass those in Louisiana.

There is also leukemia, in clusters that have suggested an environmental link. As one researcher expressed it, "I think it's a time bomb. We've got all these things floating around and are only beginning to feel the effects of them." A dissident chemist from EPA was more dramatic still. "What you've got down there," he said, "is a disaster."

As the nation's most prolific petroleum producer (49.4 *billion* barrels since 1935), and also the leading manufacturer of chemicals (with shipping orders in 1982 alone of $25 billion, or double Louisiana's total), Texas nearly certainly causes more industrial air pollution than any of the other forty-nine states. It contributes to high levels in faraway places such as the Ohio Valley.

It is like a huge pot of ozone, the hydrocarbons oscillating back and forth between Texan cities before moving north.

In their wake are the growing clusters of cancer.

One locus for such concern is Port Neches, which is about sixty miles northeast of Texas City and about eighty miles east of Houston in the Golden Triangle. It is near where the famous Spindletop gusher, struck in 1901, kicked off the modern oil industry. It is also where a local attorney named William E. Townsley has been compiling evidence of a leukemia outbreak.

The cluster is supposedly in the marshy landscape that includes Ameripol Synpol Company, a division of Uniroyal Goodrich Company and described as perhaps the largest synthetic rubber complex in the world.

Included in Townsley's research is the fact that besides a client of his who died of the disease, three others who'd attended the nearby high school between 1964 and 1974 also had succumbed to leukemia. (Another leukemia victim, according to a press report, was the seven-year-old daughter of the Port Neches mayor, Gary Graham.)

When the Texas Department of Health studied the school—deciding to handle the matter by itself instead of inviting in the CDC and its cancer experts, who were quite interested in such an investigation and would have had much less of an industrial axe to grind, not to mention much greater expertise in the research—the state found a leukemia risk no greater than anywhere else.

The study focused solely on the school's *female* population, when it was the male students coming down with the disease.

At a rate of 12.3 per 100,000 during one recent ten-year period, Jefferson County, where Port Neches is located, was more than 30 percent above the national incidence for white males. Moreover, the type that killed the high-school students, myelogenous leukemia, is one that has been directly associated with petrochemicals.

In area 5.5 times the size of Louisiana and 33 times that of New Jersey, Texas has plenty of room to spread its populace away from the poisonous fog of its eastern coast, and perhaps as a kindly result, its cancer mortality rate for 1986 was lower than forty-three other states.

But there are signs that Texas might one day challenge both Louisiana and New Jersey for leadership in particular cancer rates. In some cases it is doing so already. With the incidence of lung cancer deaths in white males at 106.2 per 100,000, Chambers County, from 1970 to 1979, was higher than the very worst parts of south Louisiana.

That is right on Galveston Bay, and the rate compared to a national one of 64. Overall, from the 1950s to the 1970s, Texas saw a 146 percent increase in respiratory cancer mortality, and as for the other kinds of cancer, they showed up like clockwork in the industrial counties one would expect them to.

There were about a thousand deaths from leukemia every year and thirteen hundred from pancreatic cancer.

In recent years Texas has been the focus of several major inquiries concerning the possibility of excess cancer at facilities such as Dow Chemical in Freeport, Texaco and Gulf in Port Arthur, Mobil in Beaumont, Ethyl Corporation in Pasadena, and the Union Carbide facility in Texas City.

Much of the attention has been on brain cancers, which are rare enough to stand out against the crowded backdrop of other kinds. There was also a case on the Houston Ship Channel where it was claimed workers at a Velsicol pesticide factory suffered severe neurological damage.

When a government entity such as OSHA has looked into brain cancers, it has found one thing, and corporate studies have found quite another. The corporations, of course, usually find no excesses, or detect them at only "insignificant" levels. ("And here we had precisely the same raw data!" says a former OSHA officer. "They found brain cancers but figured out ways of not counting them.")

While these were seemingly occupational exposures, the trends of illness are now being spotted in a way that also points to the general atmosphere.

My first introduction to Texas air had come during a trip to Lamar University years before, at which time I heard coeds complaining, like my source in New Orleans, that their nylon hose were disintegrating.

There was also an account of acid mist giving people sunburn.

The campus is so close to the industrial complex that rail cars roll and bang along the football stadium's bleachers; a reactor blows steam just beyond the fence; a shaft pipe looks like it is in the end zone.

Closest to the school is the Olin Corporation, but Mobil Chemical and PPG Industries are also within a three-mile radius.

In one study benzene was found on campus at the extremely high level of five parts per million—or 5 million parts per trillion, higher than a proposed new limit for *workplace* exposure and a level at which excessive leukemia, according to the literature, might well occur.

There are tamale stands and a shanty town in the area, along with Martin Luther King Boulevard.

There had also been a boulevard by that name in Texas City, two blocks from Rachel Chargois's house. And in St. Louis. In fact wherever there is a Martin Luther King Boulevard, it seems, factories and serious pollution are also to be found.

In 1978 the EPA very quietly initiated tests of the air in the Beaumont area and found, in one set of samples, 129 industrial-type chemicals. (It also found 210 around Lake Charles, back in Louisianian territory.) Including as they did trichloroethylene, carbon tetrachloride, and chloroform, they were exactly the kind of findings that would have caused a public hue and cry if they had been of a drinking-water source.

But the results were announced in a way that would hardly command such concern. In fact in 1986, when they were asked about the federal tests, local environmental authorities greeted the findings with surprise.

The state air experts themselves had been kept in the dark.

It seems safe to say that when it comes to publicizing air samples, the EPA prefers minimal fanfare.

Bad news it was. According to an EPA chemist involved with the tests, the agency found "basically the same general type of chemical pollution" in both Texas and Louisiana. The pollution included "many, many unidentified compounds" and some "very unusual, long hydrocarbons" that in certain instances contained twenty carbons and forty hydrogen atoms.

It can never be repeated enough how lacking our knowledge is when it comes to the way these compounds may be reacting together. If Texas

175

is indeed like Louisiana, there are compounds being formed that are not simply unidentified but do not even have laboratory code numbers yet.

Such were once found in the Mississippi River: freaks of chemistry about which we know nothing.

In an excellent study called *Respirable Particles* (Ballinger), focusing upon the extremely fine particles which are largely ignored by antiquated government regulations that are geared more toward the classic, visible soot, scientists Frederica P. Perera and A. Karim Ahmed write, "Once released into the atmosphere, hydrocarbons may be oxidized to so-called 'oxygenates,' such as aldehydes, ketones, alcohols, ethers, esters, acids, phenols, or peroxides.

"Experiments have shown that in the presence of light, sulfur dioxide, nitrogen oxides, and metallic oxides (lead, vanadium, etc.), both the oxidation of vapor-phase hydrocarbons and the formation of aerosols are enhanced."

The testing of east Texas air had caused some friction between the regional EPA office in Dallas, which bore direct responsibility for Beaumont and seemed in no hurry whatsoever for such sampling, and the national staff, which was excited by the prospects of such tests—believing, as researchers at purely academic (as opposed to political) levels of EPA did, that the Gulf Coast was indeed succeeding New Jersey as the nation's cancer hotbed.

One source told me the Dallas office gave scientists from headquarters more resistance—trying to keep them away—than the industries did.

"Most shocking has been the failure of EPA to devote resources to get the evidence they say they need before they can say something is a risk," an assistant attorney general, John Sheppard, had said back in Louisiana. "It's an incredible story because here you have the pollutants going directly from the plants to the lungs of the people. And yet it's completely virgin turf as far as research."

With the exception of southern California, the ozone in Houston, in part formed from volatile organic compounds, has been worse than about any of the nation's largest cities.

Yet Houston's chief of air quality control, Dallas Evans, a friendly man who looked like a librarian behind stacks of permits, didn't seem to think there was much of a problem at all. The industrialists are a "dedicated bunch," he said, and as a result, air quality in Houston is actually quite good—a "seven or eight" on a scale of one to ten.

"Our problems are no different than you'd find anywhere else," he

176

was somehow able to tell a visitor who had visited just about "any-where" and had not encountered anything more overpowering than what Texas wrought.

"I live here and I don't feel impacted by it," continued the official. "I don't notice any effects as a result of it.

"I don't go downtown and step over any dead bodies there."

So Texans, big in everything they do, were watching not for tiny molecules but for human corpses.

It was no wonder that in 1983 the Conservation Foundation ranked the state in its "bottom ten" for the quality of environmental controls.

(Texas was neck and neck with Louisiana for poor programs, while Alabama—which also sent some ozone across the Gulf—was ranked dead last.)

Later in the day, after my visit with Evans, I stopped at Exxon USA's headquarters downtown, in an office with a view that discreetly avoids the gassy, ether-filled ship channel.

There Jere M. Johnson, Exxon's environmental coordinator, settled into a knowing grin when I mentioned the city air chief's evaluation of Houston's pollution.

The executive himself rated the air a bit lower—"a six or a seven."

In spite of the hugeness of its operations, Exxon had done very well at avoiding environmental controversy. Its refinery at Baytown, between Houston and Beaumont, is probably the largest in the United States, closely followed by Exxon's huge facility in Baton Rouge, which I had toured before crossing state lines.

At that refinery a visitor picks up a stick-on identification badge and waits in the nicely decorated, well-secured reception area for a suave public relations manager who looks no more than forty, with the requisite smile and easy small talk, David J. Gardner.

The small talk had abruptly stopped when we sat ourselves in his office and I asked if Exxon sent any of its wastes to the Rollins incinerator for treatment.

"I thought you came here to discuss the environment," Gardner had snapped at me, his eyes suddenly ice, as if Rollins Environmental Services was irrelevant to the subject we were discussing. (One unit of the corporation had been sending waste materials there.)

But soon things smoothed out, and Gardner explained that this

177

single refinery produced 8.4 million gallons of gasoline a day—or half of what New York State would use.

With its nearby tank farms the refinery is spread over twenty-one hundred acres. By one estimate it contributes 30 percent of Baton Rouge's total industrial emissions.

But cancer at the plant? Despite one National Cancer Institute study that had shown highly excessive risks of cancer among workers in more than two hundred Texan refineries and petrochemical plants from 1947 to 1977, Exxon's own study of eighty-six hundred employees and annuitants for the period 1970 to 1977 showed a death rate 8 percent below the national average and cancer deaths 9 percent below at Baton Rouge. Working at the refinery *prolonged* one's life span, it appeared.

The important action was in upright cylinders known as pipe stills, where the crude goes in after it is heated and is separated into everything from heavy residues, at the bottom, to naphthas and light gases at the top.

Sulfur is removed by injecting hydrogen and forming the rotten-egg compound known as hydrogen sulfide.

Naphtha is the basic ingredient of gasoline but has to be broken down from large low-octane molecules to the smaller high-octane ones —meaning hydrogen atoms must be tossed off. Wielding the sledgehammers of heat and pressure, and the magic wand of catalysts (as a glossy Exxon pamphlet put it), refineries such as this can take the hydrocarbon molecules apart (or "crack" them), clean them of impurities, and put them back together again as fuels, as lubricants (such as oil and grease), or as chemical feedstocks used as building blocks for plastics and solvents.

At the end of the process, additives are blended to make the desired products: gasoline, jet fuel, kerosene, diesel, heating oil.

The cat-cracker's reactor vessel is part of a contraption thirteen stories high. But we headed for a collection of antipollution equipment that cost $350 million to build and $70 million a year to run. Key among such equipment is what is called a "wet-gas scrubber," which literally cleans the emissions by spraying water into them and precipitating out substances such as sulfur and particulates—like hosing down the smoke.

The sulfur is recovered as a bright yellow molten substance and stored in cells below the ground.

It was amazing how clean the plant itself appears, with very little spilling over.

Above us, at six hundred gallons a minute, the steam poured out a scrubber stack that removes sulfur dioxide and, in the process, was causing a steady, man-made rain to fall upon us.

Gardner was irked at how often local television cameras use the plume of steam as a backdrop while they are doing a pollution story; the stack, of course, is an *anti*pollution device.

Though there is far less sulfur coming out than used to be the case, and 85 percent less hydrocarbons, my lungs still tightened and ached dully for hours afterward, more than if I'd smoked cigarettes, apparently because I got a bit too close.

Hydrogen sulfide is still vented at ten parts per million, and as for the SO_2 (sulfur dioxide), it has been statistically shown in places like Yokkaichi, Japan, that mortality due to bronchial asthma decreases when sulfur pollution decreases. There is a clear cause and effect.

If Exxon had made impressive efforts at reducing bulk emissions of the criteria pollutants—doubtlessly saving a few bronchioles along the way—there were other companies, in both Louisiana and Texas, that often claimed an emission was just steam when in fact there was much more to it than that.

Such releases, I was told by Bill Davis, a former chemical engineer at several major companies, are virtually always a matter of either incompetence or dollars and cents.

"I've seen a number of things with chlorine that happened because of the economics of the situation," said Davis, who quit with disgust after overseeing the environmental control systems at one troubled facility.

"They figure, 'We'll let it leak until there's a more economical time to [fix] it—until we satisfy shipping orders, or until we have filled our storage capacity, or until regular maintenance comes back so we don't have to pay overtime.'

"We never discussed cancer rates. 'These coon-ass people around here smoke anyways and eat hot pepper'—that was about the only comment on cancer from the managers.

"Most [employees] are young and worried about their livelihoods. The person under forty doesn't think much about dying. You look at getting the job done at minimal costs and always meet the shipments."

Davis was himself sixty, with close-cropped hair and aviator glasses. He explained that the pollution was worse on weekends because the plants kept fewer records then, using the opportunity to purge soot from their systems—"soot-blows" to improve the equipment's heat transfers. As for toxic vapors, most releases, he said, are due to operator error.

"I've had operators who threw the wrong manual switch, an electrical knife switch, and shut down the whole plant!"

Shutting down a plant—whether by mistake or, more to the point, in order to stop a toxic leak—is risky for an ambitious young manager. "If you have too many of these things, they start looking at you with a jaundiced eye. They start questioning your judgment, or integrity, or ability to do the job.

"And your *guts*. They may start saying, 'He doesn't have the guts for the job.'

"Shutting down [for a leak] costs money. It's all based on the economics of things. Bonuses and stock options are linked to how many shipments, how little overhead—the bottom line. A bonus can be a good proportion, as much as 25 percent, of salary [for] a plant manager. Managers are judged on monthly cost reports, monthly accounting.

"If they think they'll have a better quarter a few months down the line, they may put off [a repair] until then.

"They aren't even honest with the boss. Sometimes they will downplay [a release] to the boss."

Davis also said that "if you want to evade an environmental regulation, you come up with all kinds of excuses for things happening. In most cases you can fabricate a reason. 'This happened because of a failure.' We used to fabricate things all the time. We would call the officials and tell them there was a problem at the same time we were intentionally releasing the stuff.

"It covers you in a cloak, an aura, of environmental concern. You look like a 'good neighbor.' It's easy to fool someone. We had some horrible leaks. I've seen one where we were squirting out a stream of phosgene about the size of my finger and yet chose not to shut it down.

"We reduced the pressure and after a while were able to Band-Aid it. We were always told to say, 'Oh, that's coming from somewhere else. We don't have anything going on here. We're running like a top.'

"We would actually suggest the names of other plants. 'Why don't you call them?' we'd tell them."

Since every state is currently in the process of the first major toxic inventorying and have to rely upon the factories for most of their information, I asked Davis just how accurate the companies are. When a state government asks how much of a certain compound is released, were the corporations being honest?

"Hell, no! You might know it's a thousand pounds but you might say it's ten. They're just as accurate as they feel like being.

"You knew no one could check on it.

"You tell people, the regulatory agencies, what they want to hear—unless you've just been to church or are coming out of some bar."

It was estimated that more than 650 million pounds of volatile organic chemicals were released each year by the nearly five hundred facilities in the Beaumont-Houston corridor that had government permits to release such materials.

According to corporate estimates, the Shell Chemical Company in Deer Park discharged a million pounds of benzene in 1984 (projecting no such emissions the following year); the Arco Chemical Company in Houston released 28,000 pounds of toluene (describing it as "slightly toxic" but neglecting to mention some indications that it may cause chromosome breaks, as well as potentiate other chemicals); the Mobil Corporation released 119,800 pounds of the same compound up in Beaumont; while Monsanto let go 240,000 pounds of acrylonitrile one year in Alvin. (Acrylonitrile, used to make fibers and plastics, was on a special EPA list of 403 substances that cause an "immediate threat to life and health.")

In the air downwind from two Port Arthur refineries, levels of benzene up to 405 micrograms per cubic meter were found—50 percent higher than the highest levels found initially in that part of Love Canal where black sludges were actually oozing through cellar walls. The levels seemed highest when the wind came from the direction of a Texaco facility. According to Dr. Marvin Legator, former head of the Food and Drug Administration's genetic toxicology branch (and now affiliated with the University of Texas), benzene may be more dangerous than current standards imply, and is especially insidious at low, long-term levels—perhaps *more* damaging at those levels than the high doses one would normally expect to be the most harmful ones.

"We know with benzene that there have been chromosomal effects in occupational exposure of one part per million [or 1 million parts per trillion]," he told me. "Some of our very recent studies in animals have indicated that benzene at high levels induces a certain set of enzymes which we think may detoxify benzene.

"So we're hypothesizing that at high concentrations there may be a *loss* of effectiveness [toxicity] as compared to the *lower* concentrations" (author's emphasis).

181

First isolated in 1825 from liquid condensed by compressing oil gas, benzene's major sources today include the catalytic reformate in refinery systems. Until just after World War II it was mostly produced from coal, and now more than 90 percent comes from petroleum and is used in greatest quantities for polyester, resins, and rubber, followed by such diverse products as nylon fibers, inks, and the phenols. As a solvent it helps to increase the octane rating of unleaded gasoline.

At the Mobil refinery in Beaumont, during 1983, the problem had not been benzene but an obnoxious, highly nauseating lube oil additive called dialkyldithiophosphate.

According to complaint number 102479M98 in the file of the Texas Air Control Board, which had to do with the accidental release of this additive, the subsequent behavior of Mobil personnel was almost as offensive as the chemical.

The incident had begun at about 8:40 A.M. on August 4 when the first of ninety-eight complaints arrived of an odor that was "making people physically ill." The calls seemed to be coming from across the city of Beaumont, and investigators were understandably interested in knowing what the compound was: Beaumont, after all, is no isolated backwater but a city with 120,000 residents.

Yet when a state investigator contacted Bill Stewart, an environmental coordinator at Mobil, "Bill stated that he was unaware of any problems and that they had received calls regarding the odor at about 9:00 A.M. and he had personally checked out the odor. He said that the odors were strongest over near Poplar Street and he had observed operations at Pennwalt [another company] and had observed visible emissions from their stacks. Bill said he didn't want to use names but that we should check his neighbors, that they were probably the sources of the odors."

It was the time-honored ploy of pointing a finger at someone else. Mobil hinted the problem was Pennwalt when really it was Mobil alone.

But before the refinery was going to acknowledge as much, it was going to treat the four inquiring state officials as if they were a mere nuisance—door-to-door salesmen or nosy schoolchildren. It let these public investigators scramble around pathetically, looking for the source.

When the officials sought out Mobil's Bill Stewart again to gain entry to the south part of the facility, where the fumes seemed to be coming from, the officials had to wait ten minutes for the "environmental coordinator" to coordinate an appearance at the gate—only to be told that even though he was a company coordinator, he would have to obtain permission from another superior before the state inspectors could be admitted to the suspect area.

"Although public officials are usually admitted to Mobil property even when they appear unannounced, there is no 'automatic' entry to our property or any other private property," Mobil responded in answer to my questions about the incident. "It would have been extremely irresponsible of Mobil to permit unprotected visitors to enter the immediate vicinity of the plant."

Fifteen or twenty minutes after that, Stewart finally returned to take the state people not to the trouble spot but instead to an office where another environmental specialist for Mobil, Mac Burnet, continued the stonewalling.

"Mr. Burnet would not discuss what actually happened, only that he was waiting on the final press release," reported another state official.

All the company would immediately tell the public's hired watchdogs, the public's buffer against disaster, was what was in the stingy press release.

Here is how a version of the Mobil press release in the state file read: "Early today, a slight amount of lube oil additive vapor escaped from a tank car at Mobil's Beaumont refinery. The problem was promptly located and remedied. While the fumes could be slightly nauseous near the source and smelled bad, they posed no threat to health and had no lasting effects. Cause of the accident is being investigated."

It was like describing a house fire as a smoldering mattress.

At the bottom of the curt release was a terribly vital footnote. "There is no 'e' in 'Mobil,' " it reminded reporters.

At this point, the state couldn't be blamed for wishing there *were;* a "mobile" corporation would have been one that might have moved away. That the state officials would have to scurry all around the area trying at first to hunt down the problem themselves—in the entanglement of pipes and tanks just outside of Beaumont—and then find themselves as recipients of such haughty treatment by Mobil managers, was quite enough, but what was worse was how little information Mobil could (or would) provide on the compound's toxicity. The state was worried that if the dialkyl group was configured a particular way, the highly populated area had been exposed to a compound not unlike certain pesticides and nerve gas.

"I then called Mac Burnet, Mobil, and relayed that our health-effects people needed a more specific name," reported the first investigator. "He said he would check and call me back. About two hours later,

183

Mr. Burnet called back and said their toxicologists had spoken to the toxicologist at Amoco, the manufacturer. The only further name Amoco gave was that the alkyl group was a long chain alcohol."

Mobil said the compound was not toxic. When later I asked Mobil about its description of it as "slightly nauseating," when obviously it was a bit more than that, Mobil replied, "By using the phrase 'slightly nauseating,' we meant that most affected people would feel temporary discomfort, would not become seriously ill, and would suffer no long-term effects.

"This was based on information from the manufacturer's toxicologist.

"We called the release of vapors 'slight' because we were unable to measure it. That is, so little of the additive in the tank car vaporized that there was no measurable change in the volume of additive present in the car before and after the overheating that caused the accident.

"The problem was not that Mobil lacked information on the compound in the form in which we bought it from Amoco Chemical Company—the liquid form. The problem was that no one, including Amoco, was sure of the compound's composition and toxicity in vapor form."

The state had kept pressing Mobil for more information on the chemical's toxicity. "I called Mobil, but received no answer from the Environmental Section, as they had left for the day," reported that first investigator. "I called JoAnn [a health effects specialist for the state] back and told her I would get the information in the morning as everyone had left for the day. JoAnn remarked that she certainly hoped that the company was right about the compound."

It was no wonder that on the way to Houston, motels went for $13.95 a night. Or that sticking one's hand in the ship channel there might turn the skin beet-red.

There have been problems elsewhere in the state—in San Antonio, solvents from furniture making; in Dallas, lead in children's blood; in Fort Worth, a General Dynamics plant which had released excessive hydrocarbons.

Up in Oklahoma, which is a little brother to Texas, one could find some interesting metal and benzene levels, and a pinch of whatever else was rising from the oil fields, from its zinc and chromium wastes, from the area known as Tar Creek at the Kansas border, where an environmental crisis loomed because of another moonscape of neglected mines.

Or one could go to Gore, Oklahoma, about sixty miles southeast of Tulsa, and find people like activist Jessie Deer-in-Water who claimed, near property owned by the Kerr-McGee Corporation (which operates a nuclear fuels plant there and suffered a serious leak of uranium hexafluoride in 1986), that there were badly deformed creatures in the area, including a frog with six legs and a litter of beagles with no ears. "One Indian was born with no eyes, and no sockets," she added. "She was born without eyes—not just blind. People tell me that's the type of deformities they found after Hiroshima."

Still, it is hard to compare any of that with the Houston-Beaumont-Louisiana corridor, where the stench, often, is thick as a brick.

These areas forewarned of what other areas might one day expect.

Besides cancer are the much more widespread problems of allergies and immune disorders, which could then introduce a nearly endless stream of other ailments. In Baton Rouge, teachers complained of all the children who were groggy from allergy medicine, and in the Houston area, by 1980 numbers, there were 101,002 people suffering from bronchitis and 86,956 classified as having asthma.

The excess deaths from air toxics at the base of the Superwhip alone probably exceeds the EPA cancer-death estimates for the entire nation.

In a voice tinged with Texas hyperbole, E. R. Ibert (a "coon-ass, dyed-in-the-wool Dixiecrat," to use his own description), explained that his environmental-control division in Galveston County—overseeing such spots as that of the Texas City disaster—had to make do with an annual budget of only $238,000. He also said the air regulations—"written in such a way that no one violates them very often"—had little effect on the dominating industries.

When I asked if the stuff could be traveling far, Ibert, waxing teacherlike, said that, chemically speaking, the world "is infinitely small."

The release of a quart of chemicals, if they were perfectly distributed throughout the atmosphere, would theoretically allow him to take back a quart of air anywhere in the world and pick up some of the molecules he had let out of the quart.

"You could turn loose a molecule of vinyl chloride in Texas City today, and you could have it over New York City by eight in the morning," he said, turning loose a quart of folk wisdom.

If it is impossible to take into account more than a fraction of the

185

synthetic molecules floating around the Southwest, and if there is not yet an actual body count on the streets of Houston to prove a direct cause and effect, the evidence that heavy-duty aerial chemicals are touching our innermost tissues—the innermost tissues, to some degree, of *everyone*—continues to build with relentless force.

The fact that there are no bodies on the street corners is a testimony to the human body's resilient powers.

There are hints everywhere, however, that this resiliency is in trouble.

PART IV
THE WESTERLIES

19

At the same time Texas and Louisiana were inheriting the tradition of high cancer rates from the North, out West, in California, the rates were keeping a pace just behind such places as Michigan. Though more than half the states still had a higher death rate from it (including Nebraska), California, during 1986, was expected to see ninety-two thousand new cases of cancer, or more than there were people in the city of Santa Barbara.

It is unpleasant even to hear the word *cancer,* but less pleasant still to sit back and wait for it to creep. Cancer must be viewed as an extremely avoidable cause of death, and unquestionably one way of avoiding it is to limit the carcinogens wafting about us—a proposition that not only can be, but must be, realized.

First, however, we need many more facts, and the bureaucrats in Washington have been painfully slow in culling them.

To further appreciate the questionability of government knowledge about the major air toxics, consider that if only 3 to 4 percent of the cancer deaths were attributable to them, California—never mind the Superwhip!—would by itself account for as many mortalities (1,410 to 1,880) as the EPA tried to say occurred across the nation.

Factoring in the many other types of airborne chemicals as well as people who die from chronic respiratory problems like emphysema (and not just cancer) would push the mortality figures all the higher, further emphasizing the need for an immediate reckoning of the problem—since 2.2 million Californians suffer from that kind of disease.

And not *all* the coughing is due to cars. Aside from the conventional problems of carbon monoxide, nitrogen dioxide, and sulfate aerosols,

189

which still nag badly at the state, California and the rest of the West Coast is falling victim to the more arcane contaminants that are not yet known by name to most residents of this otherwise ecologically attuned part of America.

These compounds include the chlorinateds and organophosphates, to be sure, and their unleashing into the air—whether from blowing farm particles, heavy metals, or evaporating acid—make the smog more toxic than most people think.

The situation, it should be stressed, is anything but a hopeless one. Truth is, California has shown that per capita pollution can be significantly reduced without infringing upon the economy or the free lifestyles there. Even though its population (and thus the number of cars) has grown appreciably, the standard, government-regulated pollution in Los Angeles may have improved by as much as 18 percent in just a recent five-year period, officials told news reporters.

Throughout the state, however, are hot spots of toxicity.

In the San Joaquin Valley, where pesticides are used in enormous volumes and are sprayed from airplanes which allow much of their volume to drift off target, there were reports of brain tumors, leukemia, and lymphomas from the small farming communities in places such as Fresno County.

Just west, near the Los Padres National Forest, rotten-egg odors were coming from a cotton field! The problem was a disposal well site that was being used to pump chemical residues deep below the surface there. (A local health inspector was felled by the fumes, which were also blamed for the death of an infant from lung and kidney failure.)

I recalled giving a speech in the valley and listening afterward as a woman from the audience described her own tragic exposure to chemicals. She was so weak she had to be carried backstage, where she told me about her multicolored tap water.

It was reported that three thousand wells in twenty-eight California counties had suffered contamination by poisonous, debilitating pesticides such as DDT, lindane, parathion, chlordane, 2,4,5-T, and toxaphene. These are a hint also of what might be getting into the air. Add to them, again, benzene, trichloroethylene, and the slew of other solvents, for California is the nation's fourth leading chemical producer.

Down in San Diego it isn't too bad—except when the Los Angeles smog creeps down the coast. But many other places that might appear unscathed are invisibly endangered. There are secret military installations around the state (some already known to be causing problems that go beyond the occasional puffs of missile smoke), while in other cases the bad air is generated from offshore oil rigs and comes in with what should be a refreshing, anxiously awaited sea breeze.

According to Mark Abramowitz, project director of the state's Coalition for Clean Air, one drillship used in exploratory operations can produce as much smog as twenty-five thousand cars each traveling eighteen thousand miles.

But the big concerns, of course, are on the shore, mostly in the Los Angeles vicinity but an impressive number also in the Bay Area. There is even a community thirty-five miles northeast of San Francisco named Pittsburgh.

And who was there but the Dow Chemical Company, which has produced chlorine and caustics at this plant too, and owns a bactericide facility and a plant making carbon tetrachloride.

In Richmond was a Chevron Chemical Company plant that a private research group called INFORM said released 12,220 pounds annually of organic chemicals from 161 air emission sources, many of them storage tanks. And there was also a pesticide plant there owned by Stauffer—again with an on-site incinerator, again pleading confidentiality when inquiries were made about its processes.

There are also releases of solvents and metals employed in Silicon Valley's waning, but still very significant, high-tech industry. One compound stored by manufacturers of computer chips is both highly poisonous and very flammable—a special concern in the event of an earthquake.

In one such setting, around south San Jose, state health investigators have found "an unequivocal excess" of miscarriages and birth defects, suspected of being a problem of water contamination. But once again: what is in the water, to some degree, is often in the air, too.

The valley's waterborne solvents include trichloroethane and Freon 113 escaping from a San Jose plant owned by the IBM Corporation, according to the EPA's regional office.

In San Francisco itself are any number of poisons riding the wind, some very serious in nature: during the explosion of an electrical transformer that spewed PCB-tinged smoke into a high-rise, furans had been formed.

According to the Citizens for a Better Environment, which surveyed fourteen types of toxic emissions in the Bay Area, 14.6 million pounds had been emitted in Contra Costa County during 1985 alone, including 3.5 million pounds of formaldehyde and about 1.2 million pounds of methylene chloride.

Still, what rolls in with the San Francisco fog is knee-high to the smoke and fumes blanketing the Los Angeles Basin and its 11 million residents. In the mornings, as lawn sprinklers automatically come to life —amid lilies and marguerite—the smog might start as an angry tinge of purplish red or just a smear of off-white, the sting much sharper than in a city such as New York, a bit stronger than Houston too, by noonday thick enough to obscure buildings a short distance away.

Most remarkable is that movie stars have paid such huge prices for homes with views that consist most summer days of little more than a blur of simmering hydrocarbons.

Then again, the smog and "smist" (as some call the aerosol) make for brilliant sunsets.

About 55 percent of the smog comes from automobiles, which in addition to carbon monoxide and sulfates also emit the lesser-known compounds we have focused upon: benzene, toluene, acetaldehyde, xylene, and ethylene dibromide, to name a few. That's right: automobiles are not only the major criteria source but a very significant source too of air *toxics*.

Tiny, tiny quantities of furans may also be around, with other products of incomplete combustion.

The other 45 percent comes from a list of sources far too diverse to list here. Significant among them, however, are the power utilities and oil refineries. In El Segundo, which is wedged between the Los Angeles International Airport and Manhattan Beach, is another Chevron refinery. Unlike the unabashed grind of heavy industry in Texas, here there is an attempt—however futile—to hide the clouds of steam behind hedges and flowered berms.

There is also the plume of a power plant in the area, and a waste treatment plant set alongside the surfers and hang gliders. At one time there were rumors further up the beach, in Santa Monica, of lifeguards contracting cancer. A refinery-like aroma hangs on Inglewood.

Abramowitz said that some of the highest ozone readings taken in recent years anywhere in the United States have been at a station in Glendora, to the east of Los Angeles. "In some coastal areas the fog has an acidity—they've found 1.69 [pH]—similar to that of some toilet-bowl cleansers."

Though Los Angeles is the nation's second largest city, it is known not for skyscrapers but instead for how far it sprawls.

Its smog sprawls even further. The pollution is a major problem all the way out to Riverside, and has caused visibility problems at air bases in the eastern part of the state. Sixty to ninety miles away are mountain ranges where pine needles are malformed by the pollution, and a curtain of brown smog with pollutants from as far away as both San Francisco and Los Angeles has been known to drift up the mountains into Sequoia National Park.

Even the "exclusive" area of Palm Springs frequently violates federal air standards, said Sabrina Schiller, a board member of SCAQMD, the South Coast Air Quality Management District.

Quite apropos, Mrs. Schiller is a former actress. She had made appearances on such shows as *I Dream of Jeanie, Star Trek, Mannix,* and *Hawaii Five O.* As a young girl in Ohio, she had liked watching clouds form in the sky, enjoying a clear day more than an average kid would; and about twenty years ago she had read an item about how carbon monoxide could cause coronary problems on the freeways.

Her husband is a writer/producer (*Archie Bunker Show, I Love Lucy, Maude*), and she was motivated into activism by visions of him slumped over the wheel after trying to make it through traffic from the studio.

Now Sabrina is one of the city's leading environmentalists, foe of hydrocarbons, anathema also to several fellow members of the SCAQMD board. They were trying to pressure her off the agency because she was "stirring up the troops"—too critical, too unbending, and much too willing to go to the public and tell them what was or was not going on.

To the great discomfort of local politicians, Sabrina named those who voted against smog-control efforts, parading their names out at meetings. She wanted the air clean and nothing less. It was the same kind of bold, refreshing, and yet probably unrealistic vision upon which they make movies.

Even her husband would have had problems finding comedy in this. She said vinyl chloride had once gassed a freeway, and a proposed tire-burning project might cause an extra sixteen deaths per million people —if Sabrina could believe those risk assessment methods.

As we brunched at the Riviera Country Club in Pacific Palisades, she pointed beyond the azaleas and a sunken tennis court. "We can see the improvement in criteria pollutants, but the toxic ones are very serious. What's going on here is unbelievable. Over there are the mountains, but no way you'll see them today."

Indeed, a land of lotus magic: one day a mountain range is there, the next day it is as if it had never existed. At the American Lung Association of California, Gladys Meade, environmental health director of the organization and also known as "Mrs. Smog" for her nearly two decades of antipollution work, remembered how a friend's children ran up to their mother one clear day, when the Santa Ana winds came from the desert, blowing the smog out over the ocean, and exclaimed, "Oh, mommy, the mountains came back!"

The San Fernando Valley is smothered by air so thick it looks like the fog over San Francisco Bay. When I asked if she advised jogging in the smoldering basin, Mrs. Meade was quite brief and emphatic: "Oh, no."

Some days the contrast between the fresh, clear layer of air above and the polluted air below forms a very obvious line—the strata of an air inversion.

These can occur 340 days a year and involve a layer of warm air that acts as a ceiling, preventing upward mixing and atmospheric dilution. The stagnation is made all the worse by low winds, high ozone, and the surrounding mountains, which trap pollutants.

The ozone is what is most relevant to us. Besides being the largest fraction of photochemical smog, it often indicates the presence of volatile toxic hydrocarbons such as benzene, toluene, and the dry cleaner's perchloroethylene. Ozone itself is basically an unusual form of oxygen (O_3) formed in a very complex process involving reactions between sunlight, the hydrocarbons, and oxides of nitrogen. It is not to be confused with the more naturally occurring layer of ozone that protects the earth from radiation up six to thirty miles in the stratosphere.

While officials have long emphasized the role of automobiles in supplying the necessary hydrocarbons, just as important, in Los Angeles, is the general use of such materials as paint and ink—the "area"-type contributions derived of everything from lacquer remover to car coaters to the benzene from a gas station. Coupled with these are the organic compounds vaporizing from the refineries.

In addition to the long-recognized problem of causing headaches, breathing difficulties, and reddening of the eyes, ozone recently has been shown to create a transformation in hamster and mouse cells that is

194

similar to cancer. While officials often refer to them as "total hydrocarbons," they rarely explain what specific compounds make up the ozone.

Any list would be a long one: hexanes and pentanes and butane and ethylene. The term *hydrocarbon* has a more mellow, less alarming vibration than something like "n-propylbenzene."

But in reality, extracts of particulate organic matter at the San Diego Freeway (I-405) in western Los Angeles reveal no reason for taking the mix of hydrocarbons and other materials at all complacently. It was found that these particles were both small enough to penetrate deep into lung tissue and were mutagens. They not only hurt the lungs but alter genetic material.

The problem, again, is nationwide. From 1983 to 1985, fifty larger cities and thirty smaller metropolises or rural areas did not meet federal ozone standards. But of course the Los Angeles area (most especially Riverside and San Bernardino) was the worst of them. A peak reading can reach 0.30 parts per million or more while the federal standard is 0.12.

At the same time many of the hydrocarbons forming ozone may be innocuous in and of themselves (in some places, for example, they may even come from pine trees), ozone—because it is descended, to some degree, from toxic solvents—is also, to some degree, a shadow of North America's toxic cloud.

And if some of the hydrocarbons are the result of incomplete combustion, one is tempted to ask: Should other such products of incomplete combustion, namely, the ever-worrisome furans and dioxins, be looked for too?

There are very good odds that whatever anyone wants to find would be somewhere in the Los Angeles Basin.

Though California's approach to environmental protection is perhaps more progressive than any other state, with laboratory facilities and legislation that puts it ten years or more ahead of Texas, this is not to declare that California is without instances of inadequate record-keeping, lenient enforcement, lax monitoring, and a simple lack of knowledge.

Like everywhere else, there is no statewide system of detection for the more exotic molecules.

For the four counties it covers, however, the SCAQMD did know —or claimed to know—that 15.9 million pounds of benzene (mostly from cars) are released each year as well as 28.6 million pounds of methylene chloride.

195

According to a table crammed full of digits in the far back of a report circulated in 1982 by the Stanford Research Institute, the level of methylene chloride in the air of Riverside was 0.002 parts per million, or more than four times what was found in St. Louis.

In Los Angeles, according to a more recent report, the mean level for benzene was a hair above Houston and 20 percent higher than the Lake Calumet area of southeast Chicago.

As for the accidental discharges, one particularly anguishing event had taken place at the Fiberite West Coast Corporation in Orange, where liver damage and many lesser symptoms were reported after a noxious cloud engulfed a neighborhood that included a preschool unit, a day-care center, and—as if to reach as far back into childhood development as possible—a school with a program for pregnant mothers.

The danger once more sprang from the manufacture of plastics— plastics used in racquetball rackets and the interior paneling of airplanes.

The cloud came from a batch reaction that sent fumes through the roof vents, and to its great shame, Fiberite initially withheld the specific identity of the compounds that were involved in the reaction, preventing parents from immediately knowing what their children—who were suffering throbbing heads and bleeding nostrils—had been exposed to.

Babies developed laryngitis and cried in eerie silence.

Nor could there have been much comfort in learning that inside one of the classrooms was a pet rabbit named "Peter" that had gone into convulsions and died a couple weeks after the chemical release.

One woman wore a custom-made T-shirt with stenciling that asked, "WHAT KILLED PETER RABBIT?"

According to an autopsy report by the veterinary laboratory at the California Department of Food and Agriculture, Peter Rabbit's pleural cavities had filled with a bloody fluid and the vessels in its trachea were engorged. "There is marked pulmonary edema," said the report, meaning abnormal swelling.

We'll see this type of ailment again very shortly, so keep it in mind —"pulmonary edema."

Fiberite explained its initial secrecy by saying such information was proprietary information—the old, overused, and unimaginative excuse of corporate "confidentiality."

Analyzing the decomposition products of the materials involved, the SCAQMD, according to internal memoranda, suspected that, yes, furans (of the methyl-benzo and methyl-dihydrobenzo types) may have been emitted.

Months later a pregnant mother exposed to the odors gave birth to a child with arms half the normal length and webbed fingers.

Elsewhere, a smoldering pesticide warehouse in Anaheim, near Disneyland, had caused the evacuation of ten thousand people in 1985. At first officials thought they were only dealing with a few chemicals but quickly revised their estimate to "seventy to eighty" of them.

Just days later, a couple thousand other people were evacuated from the vicinity of another pesticide warehouse, this blaze in Coachella.

The signs of a city that lives by petroleum are evident not only at the gas pumps (which are now fitted with special nozzles to prevent gases from escaping) but also in the oil wastes underground. In one case methane came flaming through the cracks of a sidewalk.

Across the street from the marbled lobby of the American Lung Association are the La Brea tar pits, where prehistoric monsters were trapped for the ages in a pool of natural asphalt. In the parking lot next to the Lung Association, more tar was oozing to the surface under a sign advertising Chrysler's XT (the "California Screamer").

To see what other toxic cases were on record in the basin, I went to the district attorney's office in downtown Los Angeles, a turnstile of hustle and bustle, and then the SCAQMD headquarters in El Monte, treading the overlapping freeways that pass by borough after borough, city after connected city—all seeming to be mere outpourings of Los Angeles.

Formerly used as corporate offices, the SCAQMD building still has a lush feel to it—thick rugs, wood tables, thriving plants. The emblem for the district is a bird in flight—too peaceful a symbol once I entered the Episode Center amid ringing, beeping phones, computer printers, and two-way transmitters that put me in mind of a high-tech police station.

Things were busy. The smog was reaching a first-stage alert.

When I asked about the unregulated type of air contaminants, not the standard smog, James M. Lents, acting executive director of the SCAQMD, said, "There are no clear guidelines on acceptable risk. Seventy to 80 percent of the effort still goes into criteria pollutants. But toxics are rapidly escalating. In a few years we may be putting more effort into toxics than criteria pollutants. Not only do you have thousands of them emitted, but they re-form in the atmosphere. If you wanted to talk on a molecular level, there would be hundreds of thousands of them."

They can derive from some sixty thousand stationary emission points along with the eight million automobiles and trucks in the basin, which are such a huge contributor that Los Angeles's air problems may not be solved until the advent of electric cars, said Lents.

As far as the point sources go, the Los Angeles district attorney's office was lobbying (and indeed later won passage) for a bill that would raise the maximum fine for intentional releases to twenty-five thousand dollars a day. "If Bhopal happened here," noted Diana Bell, a deputy district attorney, "the maximum fine right now would be one thousand dollars."

As the SCAQMD had found out, a thousand dollars a day is an easy write-off for a large company. For a limited time, so might be a twenty-five thousand dollar fine. The SCAQMD saw this demonstrated in a case involving the Mobil Oil Corporation, which has a refinery between Santa Monica and Long Beach at a place called Torrance.

The refinery seemed to have taken lessons in effrontery from its sister operation in Beaumont, Texas. It was the centerpiece of perhaps the hottest enforcement controversy in the district's nine-year history. Despite gratefulness for Mobil's annual $3 million contribution to the small city's tax coffers, there had been a number of incidents that had neighbors howling near the refinery, the flare giving off a black malodorous smoke and droplets of oil pelting the very courtroom where a Mobil case was being heard.

But the big bone was the SCAQMD's contention that the air pollution equipment at Mobil's cat-cracker—an electrostatic precipitator that charges microscopic particulates and magnetically collects them on metal plates—had been damaged and left unrepaired.

Wires had broken, plates warped, the parts removed but not replaced as Mobil went on with business as usual.

That had caused particulate emissions to exceed standards in one case by 900 percent, the district claimed. An estimated 1.3 million pounds of excess pollutants were released, described by the officials as containing lead, nickel, silica, aluminum, and chromium.

Mobil countered that the district's results were too high (there indeed had been some errors and delays in testing) and that the particles were "non-toxic, non-carcinogenic, and non-harmful."

Able to marshal more legal firepower than the district, Mobil had outmaneuvered the agency on various technicalities and won most of the important court battles, fending off an ongoing attempt by the SCAQMD to fine the company $350,000—$1,000 a day for each day the district claimed there was a violation.

The district was appealing the decision in federal court, but even a victory and full payment of that amount would have meant that Mobil saved money in the exercise. Had the refinery shut down its equipment for repairs the cost would have been much higher: perhaps $100,000 a day in lost production.

It seemed like a cold and calculated decision "in the name of corporate greed," to quote Los Angeles Councilman Marvin Braude, and the result was that "the average citizen has been forced to inhale air pollutants," said an equally upset spokesman for the SCAQMD, "in order to line Mobil's pockets."

20

Hours north of the basin, up the wondrous Pacific Coast Highway past the sea cliffs and low-hung clouds, is the tiny, unincorporated town of Casmalia, population two to three hundred, an old-time settlement among the grazeland of wild oats and burr.

It is in Santa Barbara County, the same county where Ronald Reagan has his ranch (before this he'd had property abutting Sabrina Schiller's yard!), and it consists of little more than a school, a general store with a post office, and a small steak-house.

Vandenberg Air Force Base is nearby, so astronauts and cowboys—with no little in common—can intermingle.

At the top of a hill is a crucifix that looks at least a century or more old.

"In the creeks you'll find a lot of skeletons," said my guide, Lewis Dunn, who in reality is not the type who panned gold and kept a burro but rather is a local tile contractor—big-bellied, heaving, with a crew cut and clear honest-blue eyes.

Dunn's thumb on his left hand was so numb he said he could put a flame to it and not feel a thing. It was one of many peculiar physical symptoms in an area that remined me of the afflicted neighborhoods near Baton Rouge.

"You go to a doctor and they look at you and say, 'Gee, we see no throat infection, we see no ear infection, we see no nose infection—we don't understand what's wrong with you.' Most doctors don't listen to a patient."

We passed a protest sign designed with a skull and crossbones. He whiffed the scent entering our car: "That's the acid."

Dunn had moved his family out of Casmalia because of what was just over the hill: Casmalia Resources, another example of how a waste disposal site can embitter the air as much as the factories themselves.

But this time it was neither sprayed waste oil nor an incinerator nor an old canal. It was a "state-of-the-art" facility with a laboratory, a network of interconnected white pipes, storage tanks, and a series of neat geometric pits which had been cut into the otherwise effulgent, golden foothills not far from the coastline.

It also has its own public relations office.

Into the pits had been dumped tons of winery dregs, rocket fuel, Nair hair-remover, rat carcasses, cyanide, oil-field wastes, nitric acid, and old, spoiled soda syrup.

It also has received wastes from an infamous dump in the Jurupa Mountains, the Stringfellow Acid Pits, which the federal government was scurrying to clean up with emergency funds.

That too had been a "state-of-the-art" dump in the previous decade.

But at Casmalia, the 250-acre "facility" has a wet-air oxidizer to thermally destroy some of the materials, an acid-alkaline neutralization unit, and forty-three surface impoundments. Once it had spread some of its wastes right onto the hills with spray rigs, to expedite the evaporation.

Tankers rumbled in to squirt their toxic loads into a pond. Some 328 million pounds were dumped there in 1983, with revenues another recent year of $40 million.

A couple years ago, said Dunn, fumes from the site had become especially noticeable—like home-permanent spray laced with acid.

And it left a metallic taste in your mouth.

The townsfolk began taking ill, he said, as if an extraterrestrial virus had landed.

Nearly everybody had the sniffles, and a good number something much worse. Lung cancer, leukemia, or severe nausea were claimed throughout the vicinity. At the site's gate was a red sign: "¡Cuidado! Zona de residuos peligrosos." There were trucks and more trucks, by some counts in excess of two hundred on certain days.

Circling through an area where migrant workers were bent over picking strawberries (and where broccoli, cauliflower, and lettuce are also grown, making one wonder about contamination by toxic fallout), Dunn recalled the time he came across a woman who was crying hysterically in the street because all her children were feeling ill. He had taken the family into his own house, away from the fumes, and he probably did it remembering that once, when he had been caught in the same predicament, his attorney had taken Dunn into *his* home to get him away from the smells.

As we continued through the rather impoverished neighborhood, Dunn, who once served as chairman of the Community Services District, bitterly denounced the bureaucrats he was sure must be taking bribes. He claimed that files on the waste site were missing from the regional EPA office.

He was nearly aching to be sued for libel, so he could prove his points in court. He told me about the German shepherd that hemorrhaged, and the cats that went into fits; about the black smoke, the green smoke, the school that had to close certain days here too: a little three-room school with twenty-two students from kindergarten to eighth grade.

"One day after the fumes were quite strong we came in and the canary was dead," he added about his own home.

He and his children ran a business they called, Dunn, Dunn, and Pop, but at one point it had looked like he would never work again. The business was going to pot. The fumes seemed to be making him lackadaisical.

"You'd have periods of euphoric feelings, like you were drugged, and loss of memory. I was driving the car one day in town and a kid was in the backseat—I've known him for years, he was with my son—and I looked in the backseat and I could not remember who the hell he was!

"I turned to my wife and said, 'Who's in the back with Robin?'

"I couldn't remember the kid! It took me forty-five minutes to come out of that one . . ."

There were also bouts of wheezing that in one notation from his physician, Dr. Steve Williams, were described as "chronic bronchitis chemically induced." But Dunn had fared far better than some of his peers. "I buried five neighbors in eighteen months," he said, shutting his eyes and checking his arithmetic. "Harold, Sally, then there was Bill, and then Charlie, and then there was Nora, and then there was Ruby—she died about four months after her husband. And then there was Ernie.

"The final straw that made us move was when my wife told me that she had a spot in her eye," he continued, referring to his own flight from Casmalia (to the nearby city of Santa Maria).

He took her to an ophthalmologist and found out that her retina was hemorrhaging, he explained.

Laser surgery had to be done the next morning, and though it hadn't yet filled up to her pupil, hemorrhaging had also begun in her other eye —a very odd set of circumstances, even among diabetics and others who suffer such problems, the doctor told them.

And then, worst of all, was the day Bill Jackson died.

"He was two years older than I am, and he kept telling me he had chest pains. I said, 'Bill, we've all had chest pains, right here in your chest.'

"And I got a call one day, my wife and his wife were down there talking and Bill was laying on the couch, and he got up . . . said he had to go to the bathroom, and walked past them and his daughter was in the bathroom so he sat on the bed and the next thing, his daughter walked out of the bedroom and he was laying on the bed and she walked through and the next thing they heard was a gurgle.

"Well, they rushed in there and there was blood on the ceiling and everywhere. He blew up inside. He literally blew up! I mean, there was blood everywhere. He was dead instantly. He had a massive pulmonary edema."

Dunn reached for the copy of two pages from a book listing the effects of toxic substances and showed me the page for "dioxane." It is a solvent that had been tracked in the air of Casmalia and is not to be confused with "dioxin."

Among other effects (most involving the liver or kidney), it mentioned that, in animals, dioxane had been known to cause "pulmonary edema and death."

It was also known to cause "malaise."

In the air near the school, I learned later, levels of phenols (in one sample 500,000 parts per trillion) and of benzene (higher in one sample than the highest reading logged for Los Angeles over a period of years) also had been found here.

Of course, Casmalia could also boast alerting levels of unidentified hydrocarbons.

The issue came to a head only after the fumes started drifting to other, more affluent, communities. Among them is a subdivision where the physician who handled many of the cases, Dr. Williams, lived. His office was conveniently located near a funeral home and a large graveyard in Santa Maria.

A young native of Texas who was a member of Mensa, the high-IQ society, and who still had an old-fashioned house-call bag, Dr. Williams was himself pondering a move away from the area. He felt out of sorts and fatigued.

Was it the pesticides being sprayed on local cropland?

Or the Casmalia site?

He and sixty-six other doctors signed an advertisement in the local newspaper that said, "WARNING! In our opinion, Casmalia toxic waste may be hazardous to your health."

Finally, doctors were coming to the people's defense!

Usually they avoid any comment whatsoever on such situations, wary of drawing any conclusion without proper and painstaking "proof." Somehow the chemical molecule has come to be regarded as equal in rights to a human: innocent until proven guilty.

But Dr. Williams and others had set about examining those who were most heavily exposed, analyzing liver functions, white-blood counts, bone marrow parameters, and urine screens—and finding significant traces of such chemicals as benzene and styrene in their patients' systems.

He wasn't sure Bill Jackson's death could be attributed to a classic definition of pulmonary edema, since edema usually doesn't produce such bleeding, but the case had indeed struck him as "extremely unusual" and Dr. Williams had urged Jackson's wife to move out of the area.

So had Dunn. "Nora, go to Arkansas, go to your daughter's, go anywhere else—*now!*" he had implored her.

But she had been intent on finishing the remodeling of her house. It was her way of forgetting her husband's gruesome death.

About a year later, at forty-nine, she too died from similar and unusual pulmonary problems.

There were so many miscarriages that "one of the obstetricians in the area said he's now opened up not only birthing classes but grieving classes," said Dr. Williams, whose crusade seemed to bring vandalism to his cars and an apparent break-in at his office.

But what he saw happening to the community outweighed personal concerns. The doctor was particularly struck by the case of a worker at Casmalia who Dr. Williams said had suffered a psychotic episode while working a double shift at the site and had mutilated himself, stabbing his own body dozens of times, in the face and upper thorax.

"He stabbed himself that many times and then he jumped in a pond —a toxic-waste pond," said the doctor, still aghast. "As to whether toxic

205

wastes can cause people to have that kind of alteration of mental status, the answer is very clear in the literature that it can. And the Russians have done some very good studies on the effects of various pollutants on the central nervous system. What they found basically was that the most sensitive parts of the brain to pollutants were the olfactory and amygdaloid areas. . . . The amygdaloid area of the brain is one of the areas of emotion."

I remembered claims of high suicide rates at Love Canal, but Dr. Williams had moved swiftly from that on to the area's birth defects, especially a condition—tetralogy of Fallot—where the blood flow through the heart is reversed. He was seeing that, too.

At the site itself, Jan Lachenmaier, the company's cordial public-relations manager, seemed confused over the entire fuss. "Speaking for myself, if I felt the facility were impacting people living two miles away, I personally would not want to be employed on the site. So I don't really feel those concerns.

"I would say that a year and a half ago, when we did have some odor problems, there was a giant leap taken in the community's mind that because there was something that they could smell they immediately made a link to 'This must be some sort of adverse health effect.'

"And I don't think that that link was valid. As far as the actual health effect they've reported, I don't have an explanation for that. We have over forty employees who are working on the site daily, they're in immediate contact with the materials, and none of them have experienced any of the types of problems that the community has reported."

Scorning scientific uncertainty and company PR, Dunn, who will never be convinced the fumes did not kill a good number of his friends —damaging his own lungs, filling his wife's eyes with blood—is likewise convinced that he was naive in believing, initially, that either a state or federal agency would rush to the rescue.

"The only one's gonna protect the people," Dunn said, "are the people themselves."

21

How often a stray molecule from Casmalia wanders the thirty or so necessary miles south and lands on the Reagan ranch is anybody's guess. There are times when winds blow toward the south, into the Santa Ynez Mountains, which is how the air would have to flow.

But the predominant winds are the westerlies—winds from the west heading east—and so the more relevant question is just how far *east* the collective mass of California's hydrocarbons, metal, and acid mists are heading.

Further than most people thought. The smog from Los Angeles funnels through at least two passes in the San Bernardino Mountains—Cajon and San Gorgonio—and unless conditions are such that it goes straight up and immediately joins the high-altitude global mix, the mass sweeps east over the desert known as Devil's Playground and on to the real devil's playground, Las Vegas, where it picks up some more ozone (Las Vegas is another city violating federal limits), and any radiation left over from atomic bomb tests, as well as particulates from regularly occurring inversions which have come to be named after an industrialized town nearby—the "Henderson Cloud."

From there the Californian mass drifts on to Arizona or Utah.

"Back in the early days," says William Malm, a research physicist for the National Park Service, "we were constructing transmissometers [an apparatus that measures visibility] and other optical monitors and testing them at the Grand Canyon, and as we sat there watching the instruments do their thing, we used to contemplate the haze in the canyon—a haze that's been around places like the Grand Canyon as long as I've been involved in a measurement program.

"We used to ask ourselves where it was coming from, if it was locally generated or whether the stuff was being transported from long distances. And back in those days we sort of felt like a plume being transported in from Los Angeles or from big urban areas that are three or four hundred miles away was virtually impossible. Back then if we would have hypothesized in public that this was happening, we would have been kicked over the rim of the canyon.

"But since then we have developed techniques—diagnostic tools—that use back-trajectory analysis, and we trace back air masses with certain characteristics: you measure aerosol characteristics, or particulate characteristics in the air masses, and then trace back where those masses came from. In many cases the air masses have their origins in southern California.

"And also the smelters in southern Arizona, the smelters in New Mexico, all these different source areas actually impact the Colorado Plateau, the plateau which has the Bryce Canyon, Canyonlands, Grand Canyon, Zion Canyon—all these national parks, at different times, depending on what the meteorological conditions are.

"The pollution from Los Angeles goes over one pass, and the pollutants from the San Joaquin and Central Valleys move down the front of the Sierra Nevadas and dump through another pass right next to the L.A. pass, and those two sources essentially combine and form one plume emanating from southern California across the Mojave Desert and then on into the rest.

"There are a few power plants in Nevada but Las Vegas is right in the pathway, and so L.A. tends to swamp the effect of Las Vegas and Nevada. We see Utah on occasion, too. We have seen the copper smelter up in Salt Lake City, we've been able to identify the impact coming from there at Grand Canyon also—when the wind is coming from the north.

"We've actually looked at impacts in Rocky Mountain National Park, Dinosaur National Monument, and in all these parks we see a southern California component to the pollution in the area."

And so if the Midwest is a melting pot, the Far West, to stretch the metaphor, is like a parboiler: a huge expanse that draws toxicants from eleven states and Canada on this side of the Continental Divide and has to dump at least some of them into the Superwhip on the other side of the Rockies, precooked and recombined.

The rest break up or mix upward and travel aloft, perhaps in a jet stream.

The potential movement of pollutants from the West Coast east was clearly demonstrated after the May 18, 1980, eruption at Mount St. Helens in Washington. Within fifty-four hours, according to Dr. Arlin J. Krueger of the Goddard Space Flight Center, the volcano's plume of sulfur dioxide had reached a point just south of Lake Michigan—riding air way up at the top of the troposphere.

"But some of it," he says, "must have fallen out."

And so it seems only logical that industrial particulates and vapors too—and not just those associated with acid rain—would end up a long ways from home. And that they too would fall out.

Most are fine particulates that can remain suspended for days or weeks or months.

But could the quantities really be anything to worry about?

It is the common wisdom of regulatory agencies that once you get a mile or two away from a source of pollution, levels in the air drop off to just about nothing—or at least to nothing significant. And there is a good deal of truth to that. In Western Europe, tests have shown the air in a community close to an industrial facility to have levels of benzene only a thirtieth of what was in the factory air itself.

Just a little further away the levels might be a hundredth or a thousandth or a billionth of that.

On the other hand, when dealing with carcinogens and chemicals that can affect the genetic material of a cell, it is important to keep three points in mind.

First, since we do not understand how cancer is created, and since it is conceivable that a single molecule could trigger a cell's wild proliferation into cancer, no level of a carcinogen can be considered truly safe. There is really no such thing as a threshold level.

Second, many of the compounds we have met thus far—especially chlorinateds and metals—accumulate in the ecosystem instead of being destroyed by water, bacteria, and sunlight. They keep biomagnifying— piling up.

Every little bit counts.

The third point has to do with the finiteness of the atmosphere itself. It may seem endless to us, but just five miles up the air is so thin we would suffocate for lack of oxygen.

There simply isn't that much of it to spare.

209

Though it has been reported that a compound like benzene is readily biodegradable, others, such as PCBs and DDT, are found in both polar regions. In other words, the biosphere retains many of our synthetic molecules like a big sink, swishing them back and forth, even to Antarctica.

So the final point is this: Those chemicals coming from California and the rest of the West have to be viewed with the understanding that while one city might have no effect on another in the same state, or even on one of its own suburbs, there are other times, as Dr. Malm indicated, when they *do* impact upon not only the nearest cities but upon other states and perhaps other *regions* as well.

Except for California, there is no western state that ranks in the top ten chemical producers. Nor, in most mountain terrain or expanses of desert, is there a problem with automobile emissions or refineries. There is always something to say about pesticide use (even way out in Hawaii, which uses more per capita than anywhere else), and there are always coal-fired boilers and power plants, with the concomitant benzo-(a)-pyrene.

As for accidents, toxic fumes once even caused the evacuation of a Vegas casino.

But the industry dwindles quickly once one is away from a place such as Denver or Tucson or Puget Sound.

Perhaps as a result, this block of states, from Alaska, Washington, and Idaho to Colorado and New Mexico, represents the lowest cancer rates in America. At an estimated 1986 incidence of 84 per 100,000, Alaska was the very lowest of all fifty states—a rate one third as high as Connecticut.

Yet, there are little blips of high cancer incidence throughout the rough-and-hearty territory. And other curious trends as well. According to an atlas dispensed by the U.S. Department of Health and Human Services, these western states have had higher bronchitis, emphysema, asthma, and pneumoconiosis due to silica than did most eastern states.

To think: believing it good therapy, doctors had been sending asthmatics to Arizona.

Start up in Alaska, where a urea-ammonium complex in Kenai might be the world's largest, and where a haze of oil-field smoke has changed the vistas in Prudhoe Bay; then cut down across British Columbia, past the smelters and pulping and natural-gas processing. Even lovely, mountainous Vancouver has episodes of smog.

When you get to the American border it may seem like Washington would have to be a relatively pristine state, but instead it lays claim to the Hanford nuclear reservation near Richland, which has experienced its own little renditions of Chernobyl. It is one of the most radioactive sites in the world.

In the spring of 1986 it was revealed that during the early years of its operation, routine discharges of radioactive iodine from the Hanford reservation were on the scale of what today would be considered a major accident—hundreds if not thousands of times the levels released at Three Mile Island. When documents from the Department of Energy were reviewed many years later it was found that 10 million to 100 million radioactive particles were released during certain months.

Woe to those with weak thyroids! Out at the eastern part of the state, in Spokane, meantime, there is some aluminum production going on, and roughly between Spokane and Richland is an area near Moses Lake which has reported vegetal problems as the result of a titanium operation that may have caused hydrochloric acid to form when there was moisture in the air.

It is along the coast, specifically Puget Sound, however, where the red blotches tend to show up on the atlas of illness and where the tireless cancer barometers for the period 1970–1979 show an increase over the rest of the state.

This is the Seattle-Tacoma-Olympia corridor, with some more aluminum, pulp, aircraft solvents, and oil refining (the state has four sizable refineries and about a dozen pulp mills).

Otherwise, the problems are scattered around the countryside here and in Oregon. "We had a meeting once," recalls a state official, "where someone dragged in cow lungs and dropped them on the table."

The most significant problems in the western part of America come not only from radiation but from the seemingly unlikely woodstoves. Oregon has them in droves. So does Colorado. They are more of a concern to Oregon than the small chemical firms in Portland or other of the more standard sources of hazardous materials.

The burning of wood, as natural and elemental as it might seem, is, like any other convenience, a nuisance or worse when it is improperly and overly used. In Oregon there is simply too much of a reliance on it. Besides warming houses, fire is used to clear rubbish, fungus, old grass, and weeds off timberlands and grass-seed fields—"slash burning" or "field sanitization" are the terms for this. In 1984, 237,551 acres of field in Oregon were set afire, and because some of the vegetation may have been treated with defoliants before it was burned, notes one official, "nobody's sure what comes off of it."

We do know, with confidence, that upon combustion, at least one hundred chemicals come off plain old unadulterated wood.

What could be in *wood?*

For starters it emits two hundred times the particulates that oil does, and shares certain characteristics with its organic cousin, coal.

There is carbon, of course, and when wood is burned in a stove, it is fair to expect the tarry and organic chemicals in the smoke.

About 15 percent of the nation's very fine particulates may be coming from this back-to-nature technology, according to one report.

Perhaps of greater relevance is that a fairly simple atmospheric hydrocarbon may be converted into a mutagen in the course of photooxidation. Irradiation of wood smoke can increase their mutagenic activity over that of the reactant mixture.

Furans and dioxins?

Yes, they are a possibility if a little bit of chlorine is available—for example, if the wood had been treated with preservatives or herbicides. When the Dow Chemical Company examined many common forms of combustion to see if the compound was present, it found dioxins in chimney soot and even a charcoal-broiled steak (though, in the steak, not the deathly TCDD kind).

In a controversial conclusion, the company claimed that dioxins occur from the combustion of most types of organic matter (not just from its chemicals) and thus have been around since the advent of fire.

They certainly have been around since the advent of Dow's chlorophenol production.

But be that as it may, the smoke from wood does contain other toxic components such as anthracene, phenol, and ethyl benzene. In Oregon, the first state to have instituted statewide environmental controls (in 1951), there is now a major push for new catalytic stoves that release nearly ten times less pollution *and* burn less wood. The need is an important one: in Oregon there are now 400,000 woodstoves, and 12 million nationwide.

Particulates are a problem throughout the West, especially the fine ones such as those a stove spews out. A particulate is defined as a solid particle or liquid droplet less than one hundred microns in diameter

212

(about the thickness of a hair) that can pose a threat either by mechanically damaging the respiratory system, cutting and clogging the tissue, or by acting as a carrier for adsorbed toxic substances like dioxin and benzo-(a)-pyrene.

When particulates are two microns or less in size they are especially capable of deeply infiltrating our lungs. The larger ones are caught in the nose and throat while many of those that do penetrate further—to the air ducts known as the bronchi and bronchioles, which branch off each other like limbs, branches, and twigs on a tree—are removed by tiny, broomlike hairs called cilia.

But many particles containing carcinogens like benzo-(a)-pyrene are deposited at the area of the junction where the bronchioles branch off. It is this spot, the bifurcation, interestingly enough, where tumors are most often found.

Those particulates less than half a micron across may actually make it to the deepest passages of the lung, the tiny, globular and thin-walled air sacs known as alveoli. Bunched like grapes (or leaves), the alveoli are surrounded by tiny blood vessels that exchange gases with the lung tissue, excreting such things as carbon dioxide and carrying oxygen off to the body's cells.

If a toxic gas is there, it too may make its way through the capillaries or lymph vessels.

So might a particle that is retained long enough in the alveoli.

Because organic gases more rapidly reach the blood than most particulates, and are emitted in greater volume—by one count, 162.5 trillion grams of man-made organic vapors are released in the United States each year, versus 1.6 to 16 trillion grams of particulate organics—they have tended to consume most of our attention.

Yet, particulates are extremely important as long-range carriers of toxic substances. They are thought to be partially responsible for the high level of metals in places that suffer from acid rain. Up to 47 percent of the fine particulates may also contain organic chemicals such as toluene and formaldehyde. The particulate fractions of rain samples from Portland, Oregon, have shown a long list of toxics that include such actors as dibenzofurans and phthalate esters.

Working hand in hand with atmospheric moisture, the light-scattering particulates are often responsible too for the visible pall over a city

—the "brown clouds" regularly seen in metropolises like Denver and Dallas.

In a good number of communities in Colorado the particulate levels are as high or higher than ten years ago. Anthracene and benzo-(a)-pyrene circulate through the state's thin, nearly rarefied air, trapped by inversions which, on some days, cause the air in downtown Denver to smell embarrassingly like dog urine.

As for the mountainous outbacks, wood-burned vapors float like ribbon from condominiums and ski lodges.

There is also a "brown cloud" over Albuquerque, and when the air there gets too dirty a "pollution candle" operated by the city glows with a bright red warning.

With high-tech moving into here and Phoenix, the new air mixtures, including what one official describes as "exotic new solvents," have to be reckoned with. "We roughly guessed that half of our three thousand businesses and industries emit at least one of the four hundred compounds on the EPA's hazardous list," says an official in Tucson.

If not from a vent pipe, or an open storage tank, then by accident: not only the leaks and explosions of acid mist, but outright blunders. Just east of Phoenix is the town of Globe, which, besides having a problem with defoliants, also was home to Mountain View Mobile Homes Estates, a seventeen-acre trailer park that was located on an old factory site and was so contaminated with asbestos that workers in hooded protective suits with air monitors humming on their belts had to demolish the place and bury the homes, the fences, even the blades of grass and children's toys.

About 550 miles northeast, between the cities of Denver and Brighton, is what officials with a nearly perverse relish describe as "the most contaminated square mile on earth," where the Army and Shell Chemical made nerve gas, pesticides, or other chemicals: the Rocky Mountain Arsenal. The Army had tried to pump some of the leftovers into wells deep below the surface, but in doing so caused hundreds of small earthquakes!

The region is also heavy with coal-fired furnaces and the oil industry. Hydrogen sulfide is around, and one man, in New Mexico, complained that his horse died as its trailer passed one such facility.

When I asked a Wyoming official if any people had complained of ailments near the refining up there, his reply was laconic: "Some do."

Some have to. The ozone is shifting around in heavy concentrations in some pretty unlikely places. Boulder, in the Rockies' foothills, has had ozone violations that researchers blame on the cloud coming up from Denver. The slug of hydrocarbons goes at least to the border of Wyoming.

And some technicians claim they get a pollution "signal" in the Rockies when Mexican smelters are on-line. According to Dr. Malm, the "Mexican effect"—which also includes refineries and steel mills, and the Texan border city of El Paso—has been seen too in the Grand Canyon. "We see it up in the Guadalupe Mountains, we see it in Carlsbad, in Mesa Verde. We see that influence along the whole New Mexico area. And the Gulf Coast of Texas seems to be impacting on Big Bend."

Glacier National Park in Montana feels the effects of Oregon, Washington and Canada.

There are also times, of course, when the wind blows with even greater force from east to west. It is kind of like the Superwhip snapping back behind itself. Dr. Malm says pollutants from the industrial Midwest —Minneapolis, Gary, East Chicago—have been tracked at Theodore Roosevelt National Park in North Dakota when there is a high-pressure (clockwise) system to the north or a low-pressure one to the south.

There is also a southern California "fingerprint" there, so now we're talking of transporting the junk eight hundred to eleven hundred miles.

If the fine particulates have washed out some of the Grand Canyon's crisp shadows and sun-kilned stone, the Teton Mountains can also get quite funky. Fingers point accusingly to smelters in Arizona, where unexplained clouds of ozone have hovered over the desert.

In southern Utah, where the dry desert wind whips up from Las Vegas, there is great uneasiness over the subatomic particles known as radiation.

Actually, the worry is mainly of radioactive particulates that had found their way there years before. They came from the Nevada Test Site—strontium 90 and iodine 131, from bomb explosions—and in towns like St. George and Cedar City, residents tell chilling tales of how everyone used to wake up early in the mornings and watch the sky light up like a lantern from test detonations, the fallout gathering on the cars like snow.

These "downwinders," as they call themselves, are of conservative Mormon stock—not the type to complain about something done for the nation's security. But through the years clusters of immune disorders and childhood leukemia have had them rethinking the government's atomic tests.

215

Whole families have been wiped out by cancer.

"We ran pretty much the gamut," says St. George storeowner Elmer Pickett of his own family. "Everything from leukemia to Hodgkin's disease to bone cancer to lung cancer to uterine cancer. We've had fatalities from all of those. In my family, eleven: my wife, my sister, my niece, sisters-in-law, grandmother, several uncles . . . "

There are many other sources of radiation throughout the West. I am speaking now of the metallic dust and radioactive gas (or "radon") that disperses into the atmosphere from old tailing piles and uranium mines.

The Colorado Plateau, which includes parts of New Mexico, Arizona, Utah, and of course Colorado itself, was the source of 70 percent of the uranium oxide produced in America over a thirty-year period. At the mills uranium ore has been crushed and ground and chemically separated for fuel and weaponry, its residues—or "tailings"—left in open heaps in this territory of sagebrush and soaring, golden eagles.

By one 1982 tally, Colorado, with 1,217, has the largest number of inactive mines, followed by Utah, which has 1,093. The mines may also be found in Arizona, Alaska, Washington, Idaho, Nevada, Oklahoma, Texas, Pennsylvania, Minnesota, New Jersey, South Dakota (where excess cancers are reported at the lower middle of the state), and Oregon (where the state's highest rates of breast and pancreatic cancer were found in Lakeview near 130,000 tons of mill tailings, according to a report by the Southwest Research and Information Center).

The most riveting results are not so much in cancer incidence as the reproductive effects. According to biologists Lora Mangum Shields and Alan B. Goodman, the mining and milling appear to have had a horrifying impact upon Navajos in northwestern New Mexico.

In the icy language of science, the most pronounced effects were "birth anomalies" and "fetal wastage." Shields and Goodman looked at the medical records of 13,329 Indians and found birth defects two to eight times higher than the expected incidence, including malformed hips, cleft palate, Down's syndrome, anencephaly (the severe brain defect that can include eyes that bulge because they lack sockets), and a case of cebocephaly—or "monkey face."

In Arizona the rate of birth defects was 33 percent above the national average for one period of time, and from Colorado came more reports of brain tumors and deformed skulls.

Some blamed chemicals, some blamed radiation, some blamed the highway traffic. In the case of one community, Friendly Hills, in the Denver area, near both a Martin Marietta Aerospace facility and the Rocky Flats nuclear weapons plant, reports of afflicted children arose not from state epidemiologists but from women conferring with each other at Tupperware parties.

Rocky Flats, amid the refreshing, history-rich terrain that includes Buffalo Bill's grave, had been operated by Dow Chemical and then Rockwell International, in conjunction with the U.S. Department of Energy (DOE).

The county's former health officer, Dr. Carl J. Johnson, breaking with the tradition of most such agencies, had taken an active interest in the reports of illness and began investigating the suspect winds.

Dr. Johnson's key issue was with the plutonium. One pure teaspoon of it, he was quick to remind inquirers, "would exceed the maximum permissible body burden for 140 million nuclear workers or 14 billion people."

And there were indications, he charged, that practices at Rocky Flats had been sloppy. In one case, said Dr. Johnson, a clerk was sorting and transferring files and pulled one out marked *P*.

From it fell a fragile container of plutonium-contaminated dust, which broke on the floor.

But Dr. Johnson was perhaps more perturbed that weapons-grade uranium had been found at an elementary school twelve miles south of Rocky Flats, and at another school to the east.

"This information was never reported to the public or to local public-health authorities, but had to be obtained through discovery proceedings in a large lawsuit by landowners against the DOE and its corporate contractor, Dow Chemical and Rockwell International," said Dr. Johnson, who was eventually forced out of his post, after a realtor who had been a chemical-plant manager was elected to the Board of County Commissioners.

I reached Dr. Johnson in South Dakota, where he is now state medical officer. He spoke of an explosion which had sent a black plume over Denver for half a day.

"Filters have been installed backward at Rocky Flats, permitting quantities of plutonium to escape," he told congressional investigators. "On one occasion the filters were not changed for more than four years of plant operations, until [the] explosion blew out all six hundred plus filters in the main stack."

217

(The DOE, in response, told me that only twenty-five of the filters actually blew out.)

Dr. Johnson also insisted that disrupted filters "could have contained as much as a quarter ton of very fine plutonium dust" and that the pollution "probably traveled over one hundred miles downwind." (The DOE said the amount of plutonium that escaped from the site was one mere gram.)

Rocky Mountain Sigh: This is just east of the Continental Divide, where, like a low dry breath, chinook winds take some of the other stuff from the West down the slopes and into the Superwhip.

Really, could any of the dirt make it over the Divide?

"Definitely, yes," says Dr. Julius Chang of the National Center for Atmospheric Research.

He describes the movement of ozone across the continent with a science-fiction simile: The pollution from California or Texas is like a group of astronauts heading off to the Andromeda constellation. In this case, the East Coast is Andromeda. The original voyagers would not live long enough to arrive at the destination, but their offspring would.

The same is true with ozone's precursors and products as they form and dissipate and re-form across the landscape—their progeny joining colonies of ozone on the East Coast.

Something from Texas could probably pass over Vermont or Newfoundland, says Dr. Chang, in about three or four days.

While much of the California smog that gets to the Rockies may shoot up to high altitudes, diluted beyond recognition in the global mix, other masses from the coast stay closer to the surface, depending on temperature and a myriad of other geophysical factors.

One wonders too if a minuscule, "insignificant" whiff of something from as far away as Hawaii might add a few specks to the cloud— perhaps from the burned, chemically treated stalks of old sugarcane.

Hawaii, at the same time, has encountered silicate particles from Chinese dust storms.

Are the finer particulates, the precipitates of synthetic vapor, spreading around exotic, unmonitored toxicants?

Ozone is one thing, and acid rain another. Everyone knows how far sulfates and nitrates can move.

But as for the truly nasty compounds such as chlorinated pesticides traveling at great distances, that is another question entirely. If they can make it to Andromeda, as toxaphene can, that would mean dioxins and furans and other dangerous products of incomplete combustion might too. And that would mean we all are in some jeopardy.

PART V

WHERE THE WIND CAME

22

On the other side of the midwestern vat, in the coal country of West Virginia, far from the Colorado plateau, Claire Smith's nightmare was a factory just across the road with a sign that said, "Union Carbide Agricultural Products Company."

It provoked an image that was harmlesss and even wholesome, that of wheat fields, and distant cornstalks.

But around the "agricultural" facility, she hated to tell you, were odors that caused the throat to hurt, the stomach to toss; and the complex, going for a full mile down Route 25, and in the business of making pesticides, was, as in Texas, the vast cave in which a monster slept.

It was just before noon on August 11, 1985, and Claire was herself asleep, having worked the night shift at the phone company.

What roused her was the hollering from Kirk, her eleven-year-old son. He was calling for her to answer the phone.

The voice which first seemed so irritating and vague was crystallizing into harder-edged reality.

"Girl," screamed her friend Suzette through the phone. "What are you doing still in bed? The interstate exits are sealed. They've had a leak at Union Carbide."

Claire squinted and sat up. *"People have already left!"* her friend was shrieking.

She bolted from bed and looked around. Quickly she began moving about the house. Her haze of sleep had been fully overtaken by adrenalin and confusion.

There were no sirens, nor any shouting outside, and the Carbide

223

plant was still there right out front, looking as it always looked, a giant erector set of endless pipe columns.

Inside, however, a control room had been rocked by the burst of chemicals. A solution of methylene chloride and aldicarb oxime had overheated about two hours before and burst through a safety valve.

As workers switched on the plant-wide alarm and began scurrying down hallways to escape the caustic aerosol, the cloud had seemed, at every juncture, to be a step ahead of them. The control room was engulfed. Exits were blocked by the fumes.

A couple hundred yards east, on the campus tennis courts of West Virginia State College, two professors watched a white cloud lingering near Carbide's water tower. It seemed too heavy to be steam, and it was turning gray.

They hurried out of there. Coming from near the ground, and shaped like a half-moon, it was moving toward the community.

One of the professors, Paul Nuchims, an art teacher, had painted expressionist oils of the valley's chemical sunsets, the hydrocarbons and nitrogen that regularly form a spectacular prism of odd hues. He tried to recreate them by sloshing paint and water on the canvas. In one painting, amid a flood of alizarine crimson, Nuchims had added the face of a fang-toothed monster.

It was his reaction to the disaster in Bhopal only eight months before. Those images of Bhopal! About as many people had been killed and injured in the Indian tragedy as there were residents in all of Kanawha County.

It had been the biggest industrial accident in history, 2,000 dead, 400,000 or so injured. They were gassed with methyl isocyanate, or "MIC," made from phosgene and manufactured also at the West Virginia plant.

In fact, the MIC unit was bigger here than in India!

Those images of Bhopal: rasping, blinded Indians falling in their tracks.

Or suffocating in their sleep.

There had been no refuge where the wind came.

Now, ten thousand miles west, the fears of it happening in Institute were materializing like a bad, improbable screenplay. West Virginia too was getting hit. And it was after federal inspectors had given the plant a clean bill of health. It was also after Carbide's chairman, Warren M. Anderson, in a letter to Professor Nuchims, had assured him, "We have taken steps to ensure that this already safe facility has been made even safer.''

Through Claire's window-fans came the fumes. There were always smells from the place, and the whole thirty-five-mile industrial stretch in the Kanawha Valley—a series of hardscrabble towns centered around Charleston—could most always be counted upon for one odor or another, the air stagnating most days like the inversions out in Montana and Oregon.

Gummy, syrupy substances condensed from the sky in St. Albans, and in Nitro—named after nitroglycerin—a woman complained of yellow smoke from a Monsanto incinerator.

There were orange hazes and blue hazes and in the springtime the hills were mauve.

But on August 11 the pollution was bad even by Kanawha's standards, "like the air was made of pesticides," said Claire, "like nothing I had smelled before."

Sulfur and gasoline rolled into one, with other bitter little smells, too.

Carbamates. And the fumes were pervading the house at frightening levels.

When she woke her oldest son, Kevin, he rose clutching his chest. It ached to think. It ached to breathe.

Her heart pounded. *People have already left.* It was nearly like a nuclear war!

"In panic we were running around like chickens with their heads off, not knowing what to do—turning on the radio, turning on the television, trying to get our clothes on," Claire said. "We were just running scared. I have never been afraid like this of *anything.* We heard that this toxic cloud was heading toward St. Albans, which is west of us. So what we attempted to do and had in mind to do was go *east,* to Charleston."

Peering outside, Claire, a black woman in the predominantly black, unincorporated town known as Institute, about ten miles from Charleston, searched for a look at the powdery cloud that newsmen were urgently discussing on the radio.

Framed by her window, the little community—a combination of the factory, the college, a ramshackle village—was the picture of another ghost town.

"We saw not a soul. We heard absolute silence, like we were the last people on earth.

225

"Nothing moved."

"Not even a bird," whispered her son.

No one was killed in the incident, but 135 people, including Claire Smith, were treated at the area's hospitals.

In the two days after the August 11 incident, there were also significant releases of hazardous materials in Virginia, Arizona, and New Jersey. Around West Chester, Pennsylvania, fumes from a chemical fire spread over a twenty-mile area just before the Institute incident, threatening forty thousand people.

A government report released in October of 1985 said there had been at least 6,928 accidents involving toxic chemicals in the past five years, killing at least 135 people and injuring 1,500. And this was only a partial list from selected areas. As in Bhopal and Institute, most of the problems stemmed from the storage of chemicals and human error.

The real national total is probably at least twice as high. There are fifty thousand processing units in the country that are not designed to prevent hazardous leaks during a runaway reaction. Some four billion or so tons of hazardous materials are transported each year by rail and truck—180,000 shipments a day.

In other words, there are miniature Bhopals possible nearly anyplace.

But the greater problem, again, is the routine emission of toxic substances. When congressional investigators went to West Virginia to see how much of the unregulated carcinogen acrylonitrile Union Carbide was regularly releasing into the air, they found the total in one case to be five thousand pounds per year. But then they discovered another firm in the state, Borg-Warner, of Parkersburg, was releasing nearly a million pounds a year of the very same plastic-making substance.

Small amounts of MIC have been released many times over the years. As with the majority of toxic compounds, the Clean Air Act does not regulate it.

Such surprisingly quiet emissions stand as a metaphor for the entire Southeast. Hardly thought of as a center of poison, there are, as there are everywhere, as there were even up in the Dakotas, many contributors to toxic air pollution in the region.

By virtue of its population alone, which is more dense than in states such as Arizona or Colorado, the region that centers upon Florida, Georgia, and the Carolinas, and stretches up to West Virginia and Delaware, is a significant center of both localized effects and long-distance pollution.

PCBs have been found from Gainesville, Florida, to a vault under the north portico of the White House.

On the Tarmac of Atlanta's huge airport, thick wavy fumes—benzo-(a)-pyrene-like fumes—make it difficult to breathe. So do the huge smoke clouds just west.

The awful smell of mercaptans comes from forest and paper industries throughout the South.

There are synthetic fibers from South Carolina, and plastics—some very dangerous plastics—in Maryland.

Maryland, for whatever reasons, had the highest cancer death rate in the nation (for all types combined) during the 1970s. Vinyl chloride has been tracked in the air of this state and in Delaware as well. In Baltimore, where old industrial property on the harbor is being replaced by luxury condominiums, levels of benzo-(a)-pyrene were once registered at levels higher than in the Lincoln Tunnel.

There had been spills of fumigant chemicals from the Capital Beltway to Glen Burnie. In Little Elk Valley, a hidden corner of northern Maryland, there had once been a local doctor—now taking refuge elsewhere—who had fought against a solvent-recovery plant he said had caused high levels of chemicals in the neighbors' bloodstreams, and with them excess cancer.

Airborne asbestos was a concern in Washington itself, coming, as it often did, from demolition projects.

In Pensacola, Florida, a leak from a tank car had caused a cloud of ammonia which grew to be a mile in diameter and 125 feet high. It lasted at least an hour and traveled nine miles without lifting from the ground —big enough to be spotted on airport radar.

From the navy yard at Norfolk, Virginia, come thick black puffs from both land-based boilers and the sails of stealthy submarines. In Norfolk (and also Annapolis), an antifoulant paint used on the bottoms of boats to keep off slime and barnacles has been found in the water, causing special concern because the compound tributyltin, or TBT, is toxic to clam larvae at the extraordinarily low, nearly dioxinlike parts-per-trillion level.

English scientists report that it causes the female conch to turn into a male.

Virginia has all kinds of smaller sources that quickly add up. You could take elements of about anything found in the Midwest, throw in a dash of Los Angeles, a touch of Memphis, and then condense all the gasoline emissions from one of those large, rectangular, and sparsely populated western states, and what you would come up with would be a Virginia.

The pollution is turning the Blue Ridge Mountains brown. In Caldwell County, North Carolina, residents swear the smoke from an incinerator is sometimes a sinister black or a startling crimson.

Turpentine and mercaptans seem to be all over the place, with smells that run the gamut from that of a skunk to that of orange blossoms.

Ozone reaches its highest levels in Atlanta and Birmingham, where it obscures the horizon like dusty cellophane film. In the case of Atlanta, it is often part of an inversion that can turn as red as the Georgian soil.

And it crawls up the coast. Or circulates west. At times it gets to the Mississippi and joins the mainstream Superwhip, of which it is a peripheral part.

Then, the emissions circle back northeast. The plume from Florida has been tracked in a whipping, semicircular route up through Alabama, Kentucky, Indiana, and on to New York's Adirondack Mountains.

Like the currents in the Superwhip such movements of air are greatly influenced by high-pressure systems on or near the Atlantic that often set winds in a clockwise motion.

The extent to which southern winds can push airborne material north was dramatically represented during 1982, when ash and gas from a volcano in Mexico, El Chichon, formed an acid cloud that eventually was detected over Wyoming.

From the South is where Middle Atlantic states often get their weather.

The result is a massive swish of give-and-take. Sometimes the wind blows from the north. Florida can be affected by Georgia, Alabama (where the nation's largest toxic dump is), and perhaps states as far up as Tennessee, North Carolina, and even Ohio. One Florida official says there have been enigmatic ozone readings that did not appear to have been generated anywhere locally.

This is anything but surprising considering that Florida has long been known as the inheritor of faraway dust. Its skies, especially on the east side of the state, have played host to a haze caused by dust storms in *Africa*. Driven by the wind, soil particles from the Sahara and sub-

Sahara have turned the air over Miami—four thousand miles distant—a milky, yellow gray.

Meanwhile, the emissions from such a place as West Virginia head north to New England and Canada, joining the coal-fired or chlorinated plumes from the highly industrialized Ohio Valley, which in turn can blame perhaps 20 percent of *its* pollution on what comes up from Texas.

Looping, lifting, lofting: If not to New York and New England, pollutants move due east to Virginia, Maryland, and Washington, D.C. Or revert back southward.

More relevant by far are the local releases. From just down the block, styrene or acrylonitrile. From plywood manufacturers, formaldehyde. Toxics from furniture and textile makers can be found up to Fredericksburg, Virginia.

When I visited the EPA center at North Carolina's Research Triangle Park, one of the administrators, John R. O'Connor, director of the air standards division, told me that at 25 percent of the factory fence lines, airborne chemicals are probably at higher levels than at the nation's very worst dumps. Sometimes the air is one hundred times as bad as a landfill.

Other government researchers, gathering 624 specimens of human fatty tissue from cadavers and surgical patients in six southeastern states, found that 10 percent of them contained quantifiable levels of mirex, a chlorinated pesticide that kills fire ants. Highest were samples from Georgia and Mississippi.

The estimated mean of the quantifiable amounts was 0.286 parts per million, or 286,000 parts per trillion, meaning that, were this human fat sold on the marketplace, the FDA, basing its actions on the toxic load, would have to ban its sale immediately.

When human mother's milk was studied in North Carolina, PCBs were found at levels higher than in Michigan. The study was centered around Raleigh and Durham.

One of the foremost experts in human monitoring, John Laseter of Dallas, has witnessed dozens of volatile organics and other harmful chemicals in human blood, especially the solvents toluene and xylene.

Though Dr. Laseter told me that levels in human samples from distant lands such as Australia and China are higher than in the United States (and also than England), his answer, when I asked if he ever had come across a sample of human blood that was completely *devoid* of any synthetic materials, was unhesitant:

"Never. It's not uncommon to find some of the lindane compounds there and see some of the chlordane-related materials, because they're widely used as termiticides still. Tetrachloroethylene is what's used in dry cleaning, and we'll see that in 80, 85 percent of the population— maybe more.

"I don't think we've ever seen anyone who didn't have some PCBs in the United States and North America. Virtually everybody has three or four parts per billion in their blood-free, which means that in their fat they have several orders of magnitude higher."

In Washington, D.C., Frederick W. Kutz of EPA, whose research in this most pertinent realm inspired later studies, stressed that the human levels are not just from what we swallow but from what we inhale.

"I am convinced," he said, "that air plays a more significant role in human contamination than what everyone thinks. The world thinks all the residues are from food. I don't contest that food is a route of exposure. That would be naive. But I think that we're forgetting an important route of exposure in *air*. In the alveoli of the lungs there's only a one-cell layer between the air sac and the blood system, and I think those chemicals can go through that like a jackrabbit into a hole."

And if current trends continue, more jackrabbits may be on the way. Though not yet a critical area of air toxics, high-tech is a field that bears watching. Using dozens of toxic chemicals to solder, etch, clean, and electroplate newfangled circuits, North Carolina's electrical and electronics industry has reached the point where it is producing nearly as much hazardous waste as the chemical industry.

In Chapel Hill, North Carolina, for some reason, the airborne levels of trichloroethylene were comparable to those in industrialized parts of Texas and New Jersey.

In Caldwell, at the top of Lick Mountain, was the hazardous-waste incinerator folks were complaining about. The smell wafted from Mount Hermon Methodist Church to Baxter's Store, like old honey or an overheated motor. "One child, who died of leukemia, had hair analysis showing that she had huge concentrations of lead, mercury, cadmium, selenium, and arsenic in her body," said L. C. Coonse, a high-school chemistry teacher.

Nationwide, according to an EPA computer printout, there are only 16 commercial incinerators for hazardous waste and another 235 operating privately at various plant sites where the wastes are generated to begin with.

The near future may see issuance of permits for eighty new ones.

About 3.5 million tons of the nation's hazardous waste were being burned by 1987, but with landfill space running out, and the public in an uproar over *that* manner of disposal, the agency estimates it might soon have to increase the country's volume of incineration ten times over.

As for factories, they too are a growing concern to Southerners, who no longer consider the economic benefits so sacrosanct. Though jobs are scarce and industry there still quite cherished, a random poll by city officials in Chapin, South Carolina, showed that 80 percent of those surveyed wanted to see a leaky chemical firm closed down.

Since South Carolina produces more in chemical shipments than Michigan, it is surprising there are not more toxic controversies there.

Dyes, paper, fabrics: In Atlanta, Georgia, not far from the downtown, a group spearheaded by Sister Haniyfa Ali was upset with the history of sloppy housekeeping and malfunctioning air pollution devices at a factory, AZS Chemical, that is owned by a Japanese firm and makes materials for the paint and textile industries. This time the pollution was not near a "Martin Luther King Boulevard" but near his very birthplace.

As far as pervasive and obnoxious pollution, however, virtually no industry could outdo the pulp manufacturers and papermakers who suffuse the lower part of the region with sulfur-strong, decaying smells that sting the eyes and fill the handkerchief.

In Savannah, Georgia, where one of the world's largest paper-, pulp-, and bag-making operations, Union Camp, is located, there has been a long-standing battle between a local physician, Dr. John Northrup, and the company's smelly wood digesters and evaporators.

A key concern is hydrogen sulfide. While this compound has long been known, like raw chlorine, as a killer in high amounts, hydrogen sulfide also has ramifications for those who breathe low concentrations day after day. According to a criteria paper filed in the old U.S. Department of Health, Education, and Welfare, an incident involving public exposure to less than a single part per million in Terre Haute, Indiana, during 1964 caused complaints of "nausea, vomiting, diarrhea, abdominal cramps, shortness of breath, choking, coughing, sore throat, chest pains or heaviness, headache, burning eyes . . . "

That's way less than the five hundred to one thousand parts per million that have caused unconsciousness or death in many workplace settings.

"In fact I had one fellow this morning who said that his wife got so

231

sick that he'd have to stop the car so she could get out and vomit," said Dr. Northrup. "I knew this stuff was a fantastically violent poison, a very toxic gas you could use in the gas chamber instead of cyanide."

What especially galled Dr. Northrup was how Union Camp had manipulated Savannah's collective psychology. The firm, like so many others, made the community feel indebted to it for locating there—for providing thousands of jobs—when in fact Union Camp had been near the edge of bankruptcy when the city fathers and financiers offered it low-interest loans, land at a fraction of its true value, and stipulations whereby the citizens' own money would be used to defend the company against pollution-related citizen lawsuits, according to one report.

"We are in the highest percentile for cancer, high blood pressure, cardiovascular disease, heart disease, and stroke," said the doctor. "There's something wrong in Chatham County."

The acute and long-term effects appeared also to be all around West Virginia's Kanawha Valley, too. One man, Donald Wilson, who taught near Charleston and lived in Institute, told of an exposure that caused flu-like symptoms in him and a daughter. "As the fluids developed in our lungs," he said, "we learned to sleep on an incline, head down, to keep from drowning in our own fluids."

In a deposition for Union Carbide attorneys, a woman testified that the miasma in Charleston had very ill effects on her father, who apparently had become sensitized to the chemical air. She said he would become "nauseous, his mouth begins to foam, he gets these blisters that pop in his mouth. They have burst from time to time; he has spit out pure blood, he quivers and shakes."

To the east, in a town called Alloy, where Carbide once owned another plant, the caustic air had caused an arm to fall off the torso of a Saint Anthony statue at a nearby Catholic church. It simply plopped to the ground.

The company had paid to mend the arm, and upon progressive decay (and the loss of another limb), the statue had been encased in a plastic box.

But the case was eventually infiltrated by the aerosol. And the statue had to be replaced.

At Claire Smith's, I listened to the story of another leak that took place soon after the major one. This time it was a tanker in Cross Lanes, within a couple miles of the house. When that happened her son Kevin became extremely upset.

"Everyone was still in the air over what had happened Sunday. When we got the report on the radio, on the television, my oldest son went out onto the front porch, he tried to see which way the wind was blowing. He came back in, he was shaking all over, head to toe, and he was begging me to leave: 'Momma, please let's leave. We don't know what it is and the wind's blowing this way. We could be dead sitting right here waiting to see what's happening.'

"I had been a complacent person like a lot of people around here, because I had faith in the American system and the chemical companies themselves," Claire continued. "I just felt it couldn't happen here. But after the August 11 accident I just kept thinking: It *can* happen here, and it could be as bad as India or worse. And we'd be sitting ducks.

"There was a period of time when anytime we heard anything, we were at the door and windows. And even when we heard nothing, then we would be afraid that something was going on and we were not being told and we were *still* at the doors and windows.

"When I look over there I have a sinking feeling."

She drove me to the top of a hill and explained that all the surrounding property had been owned by an ancestor of hers named Cabell (also spelled Cabble)—a wealthy plantation owner who had taken a slave as his common-law wife. Claire felt a responsibility for watching over the Cabble territory, which included the very land where Union Carbide now sat.

"Being a homeowner and the great-great granddaughter of Samuel I. Cabble, the man who owned this area, I felt bad that the people here were in jeopardy by a plant that evidently didn't care. They could've at least sent somebody to see if we were okay. They should have at least come next door, to the people adjacent to the plant. I think that angers me more than anything. My intentions now are to get out of here."

I surveyed this area where a billboard said Charlie Jones was running for state senator and Boyd Harper for the board of education. A teenager fished among the lonely reeds of a barrel-ridden backwater.

I stopped at a barbecue stand to splash water on my burning eyes. At the college library, I read news clips about how the valley, with ten thousand chemical employees, was struggling with conflicting emotions. On one side were the growing numbers who want the fumes stopped. On the opposite side were the loyal workers, who, in defiance of company critics, wore T-shirts saying, "Kiss me, I'm a Carbider."

233

When, in the wake of the August 11 release, things looked really bad, between four hundred and five hundred people took to the streets of South Charleston for a rally on the firm's behalf. *Hey, hey, what do you say; we say Carbide all the way.*

They also sent letters to the editor. It was pointed out, for example, that besides the crucial jobs it offered, Union Carbide had donated equipment and manpower in helping an adjacent rehabilitation center with construction projects. Another letter noted the many modern needs fulfilled by the chemical business: toothpaste, shampoo, cosmetics. Carbide's products were everywhere. The corporation had made Prestone antifreeze and Glad bags.

Another industrial sympathizer thought the Bhopal tragedy was something other than a case of corporate neglect: "I am a compassionate person, but it is my thinking that the leak [in Bhopal] was an act of God to put some of those people out of their misery."

One of Carbide's most strident critics, a professor named Edwin D. Hoffman, found himself awfully harassed. His tires were slashed; he got hang-up calls; things reached the point where his wife bought Mace and a cattle prod.

When I asked a research analyst at the state health department what he thought about all the controversy, and about reports of allergies throughout the area, he said, "We don't want to unduly alarm people and increase the stress-related health problems. Allergies do seem high in the area, which could be due to the pollen."

The hill country: lightning flashes beyond a foggy rise, promising more spring rain.

Below, in a night devoid of starlight, there is the MIC unit, and the watchful, waiting house of Claire Smith.

23

"I don't smell anything," said the plant manager. "Do you smell anything?"

He turned to one of his environmental specialists, a Ph.D., who shook his head. "I can't smell anything here."

A third employee, this one from corporate headquarters in Wilmington, shrugged his shoulders in agreement. He didn't detect it either.

Hear no evil, see no evil, and now smell no evil: There was nothing in the air, they informed me.

Yet I was smelling it constantly.

The odor, like husky solvents, had been very noticeable when I had first reached the E. I. du Pont de Nemours and Company's plant, known as the Chambers Works, just across the river from Wilmington, in Deepwater, New Jersey. Unless my nose was sending hallucinatory signals, the odor was still very strong and pervaded the air near the plant's administrative offices, hanging like a wet blanket. "It might be swamp gas," someone said, either the environmental specialist or Robert A. Shinn, the plant manager—a likable and patient man, a capable man, the ultimate in a company man. Fudging a bit came with corporate loyalty.

Shinn had grown up with the plant, born and raised a few hundred yards away, and he treated the Chambers Works (which is often defined as including the site of an older, adjacent facility) like his own house and car.

So here we were, in the belly of the beast; for once, inside a chemical plant looking out. For once, seeing how the vapors form.

I was being taken on a tour of the fourteen-hundred-acre site, Du

235

Pont's largest American plant and, more dauntingly, the largest in New Jersey.

Being the largest in the "Garden State" was saying a volume. This is the land of 970 chemical plants and countless chemical dumps, all crammed with abandon into a state that is the nation's fifth smallest and also the most densely populated.

At $14.2 billion in 1985 chemical sales, it now ranks second to Texas, but in the public's mind, New Jersey remains the nation's chemical capital—its dirty boy—and of course also, through no coincidence, its "Cancer Alley."

At one time, in the 1950s, the state had exceeded the national incidence of cancer by a frightful 20 percent. While the rest of the nation had come a long way in catching up since then, New Jersey, during the 1970s, had maintained a cancer rate that was about 10 percent higher than the national average and fourth highest among the fifty states (after these: Maryland, where, at 193.9 per 100,000, it appears urbanization and those area-type emissions have taken a toll; Rhode Island, which is downwind of just about everything; and Delaware, where Du Pont reigns absolutely supreme, providing not only twenty-three thousand jobs there but also one Pierre S. du Pont, who for eight years served as the state's governor).

Despite progress with other sorts of contamination, especially in cleaning some of those notorious, water-poisoning dumps, New Jersey, by 1987, still had very critical problems with its air emissions. It had become clear, says Ronald Harkov, a scientist at the state's Department of Environmental Protection (DEP), "that the so-called noncriteria pollutants may be at least as important as the criteria pollutants. The greatest health risk from the environment is from the air. The public will freak out about these compounds when they're found at trace quantities in the water, but they're a hundred times those levels in the air!"

Speaking of the same problem, Richard T. Dewling, commissioner of the DEP, told me, "We really don't understand it—we don't understand the total impact. I think it's the next era of the environment: trace contaminants in the air. Dump sites don't present the eminent-type hazard of air emissions."

New Jersey is where toxic air pollution was almost invented. According to one state official, there are about 145,000 stacks of all kinds in the Garden State.

So it was hard for any one factory to stand out from the pack. The Chambers Works, however, was up to the task. Along with a sister plant in Gibbstown, just south of Philadelphia, it produces or in some way affects 30 percent of all Du Pont's chemical products—no mean feat either, in that Du Pont is America's largest chemical company, with operations in twenty-five states and total sales nearly as high as that of all the chemical plants in New Jersey put together. The facility uses about 120 million pounds of carbon tetrachloride each year and directly or indirectly contributes $3 billion in chemical products by its many processes in Deepwater, which is at the southern end of the state, in an area that, compared to the north, is unindustrialized and even rural.

But the truck farms can be deceiving. Deepwater is the opening gate to a remarkably intense industrial corridor that runs, with a few lapses, from Chester, Pennsylvania, through Philadelphia and on to the intensely dense manufacturing just outside of New York City. There, chemical plants are clustered more closely than anywhere else in America—as common, along the New Jersey Turnpike, as all-night diners.

The Du Pont site was begun in 1891, when Francis G. du Pont was looking to expand his company's production of gunpowder. During World War I it grew at breakneck speed, from one hundred employees in 1914 to twenty-five thousand in 1918, according to Shinn, who had given me a slide show on the plant's fascinating history. "Prostitutes and moonshiners moved in," he recounted. "There were murders like the Old West, and probably the world's greatest collection of scientists."

Soon the manufacturing complex could list among its products dyes, Teflon, phosgene, petroleum additives, and the ingredients or intermediates for Dacron and also Lycra (which is used for spandex jeans), shampoo, detergents, and synthetic rubber; more than twelve hundred products had been invented or developed there, said Shinn like a proud father. "You probably have our products in your clothing, and probably our Freon in your air conditioner. Unless you live on a desert, it'd be very difficult to get more than 50 feet from a product that has been touched by this plant."

According to Shinn, *Voyager*, the plane that made it around the world without refueling, was made of materials derived from the Chambers Works.

With Shinn and his two assistants repeatedly reminding me to fasten my seat belt—and, in so doing, making sure I noticed Du Pont's safety consciousness—we set out to explore this facility which (again, with its sister plant) is responsible for some 753 end products.

In 46 million hours, since 1980, Shinn claimed, there had been no fume exposures, no cases of dermatitis, no cyanosis at the Chambers Works.

Du Pont, he said, goes so far as to advise workers on how to avoid automobile accidents and even sunburns. "All injuries, incidents, and illnesses can be prevented," said the plant manager. "We really believe that. All fume releases are preventable."

Lest anyone too quickly embrace the firm as the paragon of environmental virtue, it is helpful to know that Du Pont, for economic reasons, has taken some of the same shortcuts as smaller, less reputable firms. Like them it has trampled upon certain tapestries of nature. In the 1940s and 1950s the Chambers Works, along with other area industries, had so badly polluted the Delaware River that shad—a fish that entices many sportsmen—virtually disappeared from it.

There have been ammonia releases, chlorine releases, oleum releases. According to a report by the New York–based research group INFORM, 1.4 million pounds of methyl chloride are emitted from the Chambers Works annually, not to mention that the facility is perhaps the largest single source of hazardous waste in the nation, generating as much as the EPA once believed was produced *nationwide*. (Much of this 80 billion pounds, however, is water used for dilution.)

Once, I had investigated a leaky Du Pont landfill in Western New York that was bigger than Love Canal.

But for its size, considering the quantity of chemicals it produces, Du Pont is relatively trouble free. Unlike most other firms, it does not always try to bully officials nor stonewall reporters.

I had seen its gracious side eight years before, when, as a newspaperman, I had taken a mayonnaise jar and scooped an ugly, oily black sludge out of a polluted creek just above Niagara Falls.

My hand had turned red. Taking the sample to a private laboratory, I learned that it contained PCBs at levels the state health department said were, up to that time, unprecedented in an open water system.

There were a number of industries along the creek, and having dealt extensively with the Hooker Chemical Company, which tried pathetically to hide or soft-pedal the most basic (and to the people exposed, the most *vital*) facts about the Love Canal, I was certainly not expecting any factory to admit to the PCBs in the creek. I could not tell who it was.

But the day after my story on the pollution was printed, I received a call from Du Pont's local plant there, announcing, to my startled ears, that it, Du Pont, was accepting the blame.

Accepting blame!

Subsequently, at a cost it says was $3 million, the company dredged 12,400 cubic yards of contaminated sediment from the creek and, in a move that averted reams of bad publicity, hauled away the toxic creek bottoms.

So, though leery of the fustiness that only I seemed able to smell, I felt in fairly safe hands as the Du Ponters drove me under the pipe bridges and over the computerized rail.

It was inspiring to think that ninety thousand trucks come in here each year, and that the wastewater treatment facility, advertised by Du Pont as "state-of-the-art," could handle a city of 300,000 people. There are fourteen hundred stacks. "Twenty-six percent of all operational costs are put into environmental programs and controls," claimed the plant manager.

There was no way to verify each fact and figure, and Shinn was chock-full of them. His father, grandfather, and wife had all worked at the Chambers plant, while his mother had worked at headquarters and his son for a company operation in North Carolina.

Du Pont flowed in his family's veins like fuel in a pressurized line. There was a total of fifteen hundred years of service between his side and his wife's side, he said.

"I've had forty-two relatives on my side who worked here. My grandfather saved this town in an incident that had to do with a barge of nitroglycerin. He put out flames that had broken out on the barge by dipping his jacket in the river and beating out the fire with it."

A sturdy man with blond hair and tortoiseshell glasses, Shinn had started there as a mechanic's helper twenty-eight years before, graduating to laboratory technician, then production operator, and those years had yielded other impressive nuggets of information. For instance, Du Pont's employees or their dependents made up 58 percent of the county where the Chambers Works is located, he said, adding, more impressively, that some 222 public officials—presumably from the many hamlets and villages in the vicinity—were currently employed at the plant. The Chambers Works could boast forty-seven mayors through its history.

More to the point were the products. Once we were in the polymer-products area we were off to a unit that caused me a little trepidation: the light-blinking, sensor-laden, bell-alarmed unit of pipes where phosgene is made.

It was like a huge, steamy radiator, with upright pipes filled with a coal-like catalyst that helps combine carbon monoxide with chlorine.

Nearby were the control room and escape masks.

Used, as it had been, as a war gas, and also as an ingredient in Bhopal-like chemicals such as MIC (along with functions in fiber and dyemaking), its escape, in as little as a few pounds, would cause great public concern.

"A few ounces can kill a whole city block," fretted New Jersey Assemblyman Byron Baer, one politician who never had worked for the chemical industry.

But Shinn countered that such amounts would only be a concern near the unit itself. "Our program has been devised to have a zero release. We looked at what could be the worst disaster that could hit, like a crane crashing into it, or a plane; and the most phosgene that would be released would not be enough to get off the plant site."

A whiff can take hours to unload its damage, eventually suffocating a person in his own uncontrolled respiratory juices. And just as synthetic fibers have replaced wool and cotton, the smell of phosgene, on those rare occasions when it is smelled, is like green hay or a kind of synthetic green corn—replacing the aroma of farmland that once occupied the Chambers site.

It was as if there is a natural law in the universe that causes anything counterfeit, anything mimicking nature, anything taking the easy way out, to have certain inescapably harmful effects on us, whether acrylics to replace wool or plastic to replace wood or herbicides instead of manual labor.

But neither Du Pont nor the society it served saw it this way. Instead, the company often had been cast in heroic, herculean terms. It had provided all that gunpowder during wartime, and was just as important today as back in 1917. Who could cook anymore (not to mention run a country) without Teflon?

Its television advertisements, hyping its scientific and medical contributions, were not as banal as what Dow served up, but they were irritating nonetheless. There was the one of a bearded fellow standing next to the ocean with a Du Pont distiller and implying that the company had solved the problem of freshwater shortages with this device, which can turn seawater into tap water—neglecting to mention, of course, that shortages of fresh groundwater, especially in New Jersey or out on Long Island, were *because* of the chemical industry.

"Better things for better living," is the company slogan, and it applies to everyone who has enjoyed the luster of Du Pont paint or the cushiony feeling of synthetic mattress stuffing (which is made, in part, from deadly phosgene).

There are no "better things," however, for anyone who has inhaled too much phosgene or has just come out of radiation therapy at a New Jersey cancer ward. Most troublesome in the state have been cancers of the breast, lungs, and liver. There had also been the mysterious outbreak of leukemia up in Rutherford, next to Giants Stadium.

But what bewitched the county Du Pont was in—Salem—was an extremely high rate (17.7 per 100,000 in males from 1970 to 1979) of bladder cancer. This also ran high in counties such as Essex, Middlesex, Passaic, Monmouth, and Bergen.

Certainly chemicals played some kind of role. I learned that pilot sampling conducted right at the Chambers Works a few years before had shown the air to be filled with some of the very same types of compounds that hover over Louisiana. Said the study, known as "EPA-660/2-79-057," "One constituent of interest whose abundance was relatively large *could not be identified*" (author's emphasis).

Again, in a nearly pleading way, I longed to know what they would do in combination with one another.

"Nobody knows," admitted Shinn, just like the government people. "Nobody knows what the synergistic effects are."

As we approached the gigantic wastewater treatment facility, the odor I had smelled back at the office parking lot was getting more noticeable.

By the time we got there it was very strong indeed.

It still took a while, however, for Shinn and his colleagues to acknowledge it.

Nearly under his breath, but in a way so that I was sure to hear, Shinn blamed a company downriver.

I was a little disappointed, considering the candid way he'd handled most other questions. The smell obviously was coming from the Du Pont wastewater, and little wonder: one unit, a "primary clarifier," is the size of a football field, and enormous tanks squat nearby, bubbling air up through the liquid materials.

This is extremely important because there are plants like this— settling, aerating, and otherwise filtering and treating wastewater— everywhere in America. At least 3.2 billion gallons of water laden with chemicals and metals flow to publicly owned treatment facilities every day. Around one in Philadelphia the wastewater gave off volatile organics such as chloroform, dichloropropane, and a compound that, like furans, should get more publicity, ethylene dichloride, which is a fuel additive and, besides boasting its own substantial toxicity, can help form furans and dioxins.

There are similar releases in every major city, with problems just beginning to be identified but already being talked about in Indiana, Illinois, California, and New York. Nationally there are thirty-nine thousand treatment facilities identified as important sources of priority pollutants, and, of these, twenty-four thousand are on industrial property and fifteen thousand are publicly owned. Their fetor can match a large factory's. And from the perspective of contaminating the air, these wastewater plants, it appears, may be more significant than toxic landfills.

Clean the water and you dirty the air: There seemed no solving the toxic conundrum. But at least now the situation was being seen by corporations such as Du Pont as a major area for capital expenditures, not just as a PR problem. The public's outcry is starting to be heard. By using new techniques whereby floating roofs are placed on storage tanks to minimize the escape of gases, and whereby filters, in a boxlike unit, remove fine particulates, Du Pont has demonstrated environmental foresight and acknowledged that the public's concern—indeed, the public's *anger*—is not a fleeting mood.

"Some companies still resist the inevitable while they seek to establish a margin of advantage over their competitors. Such conduct is hurting the rest of the industry and should not be defended or tolerated," said Dr. William G. Simeral, an executive vice-president.

His corporation had learned the hard way that some compounds, while profitable in the short term, are big losers over the long haul. The best example was Du Pont's experience with beta-naphthylamine and benzidine, which caused bladder cancer in its workers.

Though Shinn felt comfortable enough with what Du Pont emitted to live within a couple miles of the Chambers Works—and had seen no ill effects on his grandfather, who lived to be ninety-three, or his father, who worked with dyes and was still alive at ninety-seven—he also had an uncle, a couple of cousins, and a father-in-law who had all succumbed to a single disease: bladder cancer.

Even an official spokesman for the state's environmental agency had suffered a cancerous condition in his urinary tract.

Where a state such as Montana might have one county in twenty-one where the cancer mortality is significantly higher than expected (and is shown as such with the asterisk), New Jersey, during the same time, would have *sixteen* of its twenty-one counties flagged.

What's more, those numbers are for white females and not for the husbands who make up most of the factory work force.

There is also reason for epidemiological concern in the part of Pennsylvania that borders New Jersey. This state has 548 chemical plants of its own. The mutagens surround Philadelphia in gritty ourskirts like Chester where steel operations and refineries turn the skies into a violent splash of flue gas.

The area is also the inevitable recipient, from time to time, of others' sullied air. At one time, according to a private report, the upwind county of New Castle, in Delaware, was emitting half as much sulfur dioxide as the city of New York.

The eastern seaboard, awash in its own outpourings, is also the final point in America where residues from the Superwhip, the Canadian Cool, and the Western Flux all converge, with an effect that is largely unsearched and thus largely unknown.

In Philadelphia proper, the cancer rate for white males during the 1970s was nearly 8 percent above Chicago's.

But the most striking hot spot on the East Coast is the black swath that opens up near Trenton and moves through the middle of New Jersey until, in the state's northeast, it balloons like the top of a Minnesota tornado.

Along the turnpike toward Newark are the famous sights and smells. Forgetting its ocean shore, its juicy tomatoes, its beautiful Pine Barrens, visitors from around the world traveling to the airport remember the rotten eggs of hydrogen sulfide and equally sickening smells that resemble burned plastic, burned mustard, and soggy vitamin capsules.

243

Orange clouds, red clouds, one that was pink: little puffs every-where, some so close to a factory's soil that they look like a brushfire.

Smoke squirts from low rooftops in Newark or plumes a thousand feet beyond a refinery.

The sky gray or brown or shimmering lime oxides.

Much of the northern part of the state is composed of rolling hills and affluent suburbs. But nobody—not in Glen Rock, not in Upper Montclair—can be totally free of the insidious molecules spreading from the many dumps. These landfills are in some cases several stories high: gull-ridden, paper-strewn mountains of trash rich in tales of buried Mafia corpses and other midnight dumping.

According to popular lore, some of the bodies were crammed into fifty-five gallon drums. Others were dissolved in acid.

With the supervision of an Essex County detective, I interviewed an informant who, driving his fur-lined-interior Mercedes up a squalid dirt road near one of the dumps, said he had participated in a number of mob killings and that the bodies ended up either in an asphalt hopper, in a local incinerator, or "would be left in the back with the chemical shit and covered with a bulldozer."

At night, through the eerie rolls of fog, lights twinkle on the risers and standpipes—on the reactors and fractionators of a refinery—like stars in a synthetic firmament.

There is actually a sign, at a liquid petroleum off-loading gate, that says, "DANGER . . . DO NOT DRIVE INTO VAPOR CLOUD."

"Twelve thousand tons of these hydrocarbons, or unburned fuel, escape into the atmosphere each year in New Jersey simply from cars fueling up at gas stations," an official complained to the Newark *Star-Ledger*. "That translates to 4.5 million gallons of gasoline."

Fumes are also forced out of tankers loading up.

A state official told me there once had been a test that showed the area near Newark International Airport to be the single most polluted spot in the nation. But it depends on what chemical you are looking for, and when the samples are taken. Benzene, in a run of samples by the New Jersey Institute of Technology, looked very high in the city of Eliz-abeth, but not as high as it was in Houston, Toronto, Phoenix, Tusca-loosa, or Los Angeles.

But only in New Jersey was there an incinerator which malfunctioned and spat discarded birth-control pills onto a nearby street.

Back downstate, in Logan Township, in the county immediately north of Salem, is another Rollins site.

Meanwhile, dioxin had been found at a public swimming pool in Newark, near an old Diamond Shamrock plant. In the cracks of the cement! In other communities nearby, near other plants, white fallout had covered lawns like an overnight snow. Workers at one area plant had developed dime-sized holes between their nostrils. One of the men could push a napkin up one nostril and pull it out of the other.

Overhead was the constant roar—and constant exhaust—from planes lifting off the international runway.

Hundreds of thousands of residents had caught whiffs of a series of recent leaks. In one snakebitten period of just four months there had been at least fifteen significant incidents, including an escape of organophosphate pesticide from American Cyanamid in exceptionally polluted Linden. Next door, at another Du Pont plant, a release of oleum had carried along Arthur Kill in an early-morning smog and overcome forty people, including workmen painting the Goethals Bridge.

These incidents were small, however, when set next to one on October 21, 1985, from the Exxon refinery in Linden—another stupendous cat-cracker surrounded by more than seven miles of fence and ball-tanks that looked like space modules. In this case an overheated reaction sent such a surge of hydrogen sulfide to the flare—which is supposed to burn such excesses—that the device was overwhelmed.

Exxon tried to call the release "small," but in reality its smell was noticed over two thirds of New Jersey and all the way down to Philadelphia. Police switchboards were as overwhelmed as the flare. From the Exxon sewers came the powerful smell of mercaptans.

As if to dispel concern about such things, officials at both federal and state levels throughout America have begun pointing to an EPA study that showed toxic levels to be higher *inside* homes than *outside* of them.

There is great merit in weighing this aspect of toxic air pollution. With all the chemical products inside a house—mothballs, nail polish, propellants, paint, plastics, cleaning agents, and gasoline fumes (if there's an attached garage)—it is only logical that New Jerseyans, and all Americans, have a serious worry too about indoor air pollution.

Consider this: When EPA scientists attached personal, portable

245

monitors to 370 people in New Jersey and also took periodic readings of their breath for certain prevalent chemicals, these scientists discovered, to their shock, that indoor exposures in the home, collected overnight, were usually two to five times (and sometimes more than seventy times) higher for eleven compounds than outdoor exposures, even for those living near the Jersey plants that *made* such compounds.

In fact, New Jerseyans, for this limited number of compounds, were only slightly more exposed to volatile toxics than people living in Greensboro, North Carolina, and Devils Lake, North Dakota.

Though such powerhouses as furans, dioxin, and benzo-(a)-pyrene were not part of the study, the solvent types are enough to warrant major concern.

Perhaps as bad is the fact that many New Jerseyans—again along with millions elsewhere—live in homes which have been infiltrated by radon, the odorless, colorless radioactive gas. Suspected of afflicting between 1 and 8 million American homes in more than thirty states, radon comes from nature itself, the decay product of radium, which in turn comes from uranium in the soil.

Some 250,000 homes in northwestern New Jersey are on top of a uranium-laden strata of granite that stretches from Reading, Pennsylvania, up to Orange, Putnam, Rockland, and Dutchess counties in New York. I spoke to one man, Stanley Watras, whose home had radon levels higher than just about any other on record, supposedly equivalent to his smoking 220 packs of cigarettes a day or, on a yearly basis, receiving 455,000 chest x-rays.

One threat, however, does not negate another. Instead they all work quite well in concert—altogether too well, a toxic cloud from within and without.

Outside, the New Jersey air, joined by any hydrocarbons from the Canadian Cool, the westerlies, and the Superwhip, moves out over the ocean or across New York City to Long Island, as well as the upstate suburbs and New England.

The strongest pulse is not simply to sea but instead northward first.

"For a long time we've used the atmosphere as a garbage can, and what goes up must come down," says Dr. Reiss of Rutgers. "We're just beginning to understand what's going on, and it could be a Pandora's Box."

24

There are little Pandora's boxes all over the mid-Atlantic territory just before the ocean.

Pennsylvania and upstate New York, for instance, have much in common with the industries of Ohio and New Jersey. Responsible at one time for 25 percent of the nation's steel production, Pennsylvania is also among four states in the lake region that rank among the top ten chemical producers—in the slot just below New York, which has 60,000 registered emission points.

This inland stretch between the Great Lakes and the Atlantic coast, including as it does Pittsburgh, was, like New Jersey and parts of New England, where serious industrial pollution first began.

The problem with understanding Pennsylvania is that its Department of Environmental Resources, or DER, is sluggish, secretive and, even in comparison to most other obsequious states, overly protective of industry.

Instead of emphasizing the benzo-(a)-pyrene that is still fairly heavy around Pittsburgh, or the hydrogen sulfide that came up an old gas bore at a state park near Erie (perhaps as a result of paper-making wastes that had been shot down a deep-well!) state officials shied away from inquirers or ignored the problems.

Once, near Scranton, I informed the DER that a Mafia-run junkyard at the top of a mountain seemed to be clandestinely handling hazardous wastes. A helicopter surveillance that I had ordered showed the ground coated with multicolored residues. It also appeared that electrical wiring had been burned up there, which could unleash the extremely dangerous residues of plastic. Neighbors had complained of headaches and nausea.

But the DER, after viewing the aerial footage, had shown only a passing interest in the suspicious, hard-to-find mountaintop, and did anything but storm the site.

In this case, since the owner of the dump was known as "Black Bear," the problem may have been simple fear: Rumor was the place also had burned some New Jersey bodies.

There are competent people in the DER, however, and they can point to some solid success. Though the Pittsburgh area still has significant problems, and emits at least 13.7 million pounds of organics a year, it is far cleaner to the eyes than in the olden days before Earth Day—before air regulations—when the ovens were running full blast and unabashed.

Back then, near one center of steel manufacture, drivers had to use headlights during daylight hours to pierce through the thick acrid smog.

Though much of the industry has left, the soot-strewn stretch between Harrisburg and Scranton still has piles of old, hard-coal rubble. Behind the DER's back, midnight dumpers had poured highly dangerous chemicals down the shaft of an old coal mine.

Up in New York the state is beginning to move swiftly into the realm of air toxics, looking not just at the factories but also smaller, yet cumulatively important, sources such as the local gas station, where levels of benzene can reach fifty thousand parts per trillion near a pump —or about ten times what would normally be found in the air of a city like Burbank, California.

By one federal estimate, gas marketing causes nearly as many cancers as chemical production does.

There is also the dry cleaner to consider. Of an estimated three thousand toxic compounds regularly liberated into the air of New York State are two—trichloroethylene and perchloroethylene—that can be found almost anywhere.

In a state as populous as New York (it ranks second to California), such small area sources add up rapidly.

Another compound that has caused trouble in upstate New York, outside and inside the home, is carbon tetrachloride, which is very difficult to break down even at high heat. By the year 2020 it is predicted that, nationwide, five hundred to twenty-two thousand annual cases of skin cancer might be caused by carbon tetrachloride (or "carbon tet," as those who are around it affectionately call the compound). More omnipresent, no doubt, is formaldehyde.

The pollution has gone long distances in unexpected concentrations.

A great horned owl found dead in Catskill contained 357 parts per million of PCBs! Up in Hamilton County—part of the Adirondack region, where hundreds of lakes are under the assault of acid—the rate of lung cancer from 1978 to 1982 was nearly *twice* that of car-cluttered, diesel-ridden Manhattan. Syracuse, Rochester, and Buffalo are home to chemical production, electronic manufacture, papermakers, photographic products, cement factories, silicon, surface coating, pharmaceuticals, and the VOC-laden printing industry. In Binghamton a state office building had to be abandoned because of contamination by a PCB fire and the resultant dioxins.

As for claims of mysterious cancer clusters, New York State has enough of them to have opened a special bureau to handle such investigations. Way east, Long Island is a hotbed of cancer fears, but the problem is fully statewide and is familiar to those near the lakes, too. In 1981 there had been sixteen inquiries from concerned community groups. In 1984 there were three times that number.

There are so many toxic dumps in the countryside that the state has passed a bond issue allocating more than $1 billion to stop them from leaking. Once a toxic chemical is in the water, the "grasshopper effect" often takes place. That is, the chemical is stirred up from sediments, travels to the surface, vaporizes, and eventually rains down there or somewhere else.

Up north of Albany were the spots on the Hudson River where General Electric discharged PCBs into the river, contaminating the fish and making PCBs a household name. Not far away, in Glens Falls, according to Donald Gower, director of the Bureau of Air Quality Surveillance, low levels of PCBs have been detected in the air. To everyone's mystification, they have also found their way into Lake Champlain.

The air: It is good to remember that while DDT and PCBs for years had been measured in soil, fish flesh, and other substances that readily accumulate them, it is not usual for them to be tracked either in running water or circulating air.

It is one thing to detect PCBs or dioxin, especially the incredibly toxic 2,3,7,8-tetra isomer, in the fish or soil of areas like Midland or Love Canal, but quite another matter if the material is tracked in the flowing atmosphere.

Nearly to their relief, environmental bureaus found dioxin enormously difficult, with current equipment, to detect in the air. California is only beginning to think of looking for it, and only a handful of laboratories could come up with reliable and indisputable techniques. It might take dozens of samples to catch a good glimpse of it, and each sample could cost more than two thousand dollars.

But New York has the technique in place, and its laboratories are more sophisticated and extensive perhaps than any except those at the CDC in Atlanta or the EPA's facilities at Research Triangle Park in North Carolina. One of the state's recent achievements, said George A. Eadon, director of environmental sciences, was the capability to isolate dioxin in the air—the flowing air!—in the parts per quadrillion.

It was like peering into a distant galaxy. At about sea level there are approximately 10 billion trillion air molecules in a breath of air, or as many molecules as there are stars in the known universe.

So one part per quadrillion, tiny as it seems, might translate into 10 million dioxin molecules in each breath!

Before any panic, one should realize there are trillions of air molecules between your eyes and this very page. Some of them may have once flowed from the lungs of a seventeenth-century man, or been part of a high-flying cloud just last week.

At such small levels you could find at least one hundred and perhaps many hundreds of compounds in a new, vinylized car.

But perhaps all it takes is one molecule of an extraordinarily reactive chemical to knock down the door to the nucleus—the control center— of a single cell, which might then begin the cancerous proliferation. Any amount of dioxin, confided another official, would be "scary."

Dr. Eadon took me down the beautifully paneled and carpeted hall outside his office and into the laboratories that consume 500,000 square feet of space in the futuristic complex of state offices known as the Empire State Plaza.

We went in and out of laboratories geared with the gas chromatagraph (basically a coil where compounds are separated out according to their boiling points and affinity for the coil), and the more recent mass spectrometer (where ions are separated in more detail by magnets or electric fields).

The chromatograph looked like a microwave oven without the door, the spectrometers a bulk of tubes and wires and pumps surrounded by wavy green computer screens.

These tools are the telescopes into the infinitely small world around us.

And on the door into one room was the sign, "Danger, Do Not Enter."

I felt like whispering, for under a special venting hood, on a table of beakers and eyedroppers, was where they handle samples of TCDD.

The first big test, said Dr. Eadon, was to see if there was dioxin in the air of one of the most industrialized parts of New York State, the city of Niagara Falls.

He explained that the laboratory worked long hours sorting through the graph peaks, trying to match them specifically—filtering out the interferences, or "noise."

There were peaks that *seemed* like dioxin, but there was much doubt. This was wholly new turf: no state agency anywhere had confidently, indisputably detected dioxin in the community air before. It would be a "breakthrough." It would be without much precedent. This was *not* soil sampling. This was not sewer sludge. Finding it in the ambient air, in that other official's description, would be "scary."

"We were really quite surprised to start seeing ambiguous signs, and then we started to get clear, defensible evidence that it was there," Dr. Eadon explained.

The chromatograph was set up to separate the compounds, the mass spectrometer to gauge their precise molecular weight. There were jagged peaks, but they were too close to the peaks that were possibly caused by other chemicals or by electronic interference.

The technicians were pushing technology to the outer bounds, and the signal-to-noise ratio at first was just too low. They waited for more samples, perfected the technique, fine-tuned its discernment.

Then, in a jolting set of peaks on the graph, his scientists unmistakably identified 2,3,7,8-tetra dioxin in the ambient, free-flowing air in the most polluted part of Niagara Falls.

Dioxin of the worst isomer was in the air, captured in cartridges containing an adsorbent gel.

It was above the detection limit of a part per quadrillion.

The neighborhood had been in the shroud of toxicants for decades, like living in a miniature combination of Pittsburgh and Baton Rouge. Years before I had spoken to thirty-two residents whose families and friends seemed to be suffering as much as the people I had once surveyed at Love Canal.

Gardens were stunted, or the leaves withered, like in Midland. Dogs got rashes, and house siding had turned dark.

When I asked one woman about how the air may have affected her, she had lifted off a hairpiece.

The state suspected the key source of the dioxin to be a waste-to-energy plant Occidental Chemical (formerly Hooker Chemical) had recently begun to operate. It is the company made infamous a few miles away, at the Love Canal.

Drama indeed. The state had found dioxin in stack tests at similar plants in Westchester and Albany counties, but those were stack tests, not evaluations of the surrounding air.

It is now however a good bet that most places scientists look, if they look hard enough, if they *want* to find it, they will similarly locate dioxin. In so doing they will present society with a new problem that we can—that we must—then quickly move forward to resolve. Foolhardy as it may sound to suggest that perhaps the most toxic compound synthesized by man is now not only in Midland or Love Canal but across the land, I base this, in part, upon an article which appeared in the November 2, 1984, issue of *Science*. It was authored by researchers from Indiana University who reported finding furans in sediments of Siskiwit Lake on Isle Royale.

Taking core samples, they traced most of the deposition as coinciding with the nationwide boom in chlorinated processes—knocking down Dow's implication that such compounds are caused by simple combustion and not so much by industrial operations. Tests of particulates from St. Louis and Washington, D.C., also proved positive.

"These results suggest that . . . [furans] are ubiquitous environmental contaminants that are transported through the atmosphere much in the same way as pesticides and certain heavy metals," wrote the authors. "We can certainly rule out natural combustion (for example, forest fires). The historical record of . . . furan deposition is preserved in sediments, and this record shows that . . . furans were virtually absent from sediments until about 1940."

Scientists speculated that along with the Dow facility in Midland, midwestern trash incinerators are contributing to the fallout in Siskiwit Lake. For down in the muck of this freshwater pool 650 miles northwest of Pittsburgh—amid the loons, amid the arboreal forest, in one of the few truly untouched preserves left in America—dioxins were also detected.

252

25

Finally, there she stood, Miss Liberty, her back to the tank farms, her new torch rising from a robe of New Jersey haze.

I studied her from an office in the World Trade Center that belonged, ironically, to the New York Department of Environmental Conservation.

Somehow, in the murk, the statue remained a strong signal of hope. This time the hope is of liberty, someday, some way, from the aging century's new carcinogens.

Before solving the problem, however, we must comprehend the extent of it, and until 1987 the sanitation commission overseeing the city of New York and vicinity did not even have regular access to the equipment needed to look for the most potent hydrocarbons.

Indeed, in this, the nation's largest, most sophisticated city, there had been only the barest spot-testing to see whether, as it is reasonable to expect, the population is inhaling furanlike molecules as it mills about Wall Street and Greenwich Village, where so much of the nation's wind arrives.

"The public has a real fear and I can't say it's unfounded," says Alan Mytelka, director of the Interstate Sanitation Commission. "We get asked, 'What's in the atmosphere?' and the answer is, 'We don't know.' "

Once, during a basic benzene study, he says, "We went out to the middle of Central Park and the numbers were higher than hell. . . . We saw other squiggles and knew something else was there. We know there are materials out there that are toxic."

253

On Staten Island, across the narrow bayway from New Jersey, within nearly a stone's throw of the industrial vents, respiratory ailments are legion and show signs of growing worse. They are so numerous, in fact, that support groups have been set up for the parents of asthmatics. At one hospital the visits had increased 20 percent during a recent six months.

Though lower than the Louisiana incidence, rates for cancer of the trachea, bronchus, and lung are higher than all but a county or two in New Jersey itself. From 1960 to 1980 the incidence had increased by 36 percent. That gives Staten Island the status of "hot spot," with a rate 10 percent higher than the rest of the city. Of the smells plaguing the island, none was more obnoxious than the pervasive one resembling cat urine.

It is like in Linden, which is just across the water: sewer smells, or sweet-sick threads of solvent.

Outside one pigment factory on the island, snow had turned a startling red.

One woman said the dust was so discolored inside her home she was going to change the name of her dog from "Whitey" to "Red." Another woman, suffering bouts of dizziness, threatened, in the blunt style of New Yorkers, to buy two attack dogs and "sic 'em" on whoever was responsible in New Jersey.

Thousands of feet above, the southern flow from Florida and Georgie might be bringing in a front of storm clouds. From the west, there might be thicker, grayer clouds mounting like a wrinkled carpet and bringing forth scattered collections of agricultural dust and combustion residues.

Or perhaps, below a 140-mile-per-hour airstream, the flow was slow and mellow.

In January a blast of Canadian Cool or a classic nor'easter might form piles of snow that soon, on Manhattan streets, would be black and doggone yellow. Variations in ambient concentrations, said a report by researchers from the New York University Medical Center, "suggest that the long-range transport of the atmospheric aerosol can also contribute to burdens of PAH (polycyclic aromatic hydrocarbons) in New York City and other regions of the northeastern U.S."

Was toxaphene from the South sweeping across Central Park?

Probably so, and PCBs almost certainly.

But the toxics arriving long-distance, we must always stress, are likely to be very tiny and more of an additive, steady-stream, and extremely subtle factor compared to the strong local emissions. During the experiments in which tracers were released from Dayton, Ohio, levels were about two orders of magnitude (one hundred times) lower in New York, Connecticut, and Vermont than they were 150 miles downwind of the release (which, by then, were already greatly dissipated: ten thousand or so times less than the concentration at the actual release point).

To the Big Apple, however, New Jersey is a local source. In fact Manhattan's western shore is a mere five air-miles from Giants Stadium.

Since the factories get denser as one moves south, and the predominant wind flow is still northeast, that means, along with the 1987 Super Bowl champions, that the nation's largest center of population, the world's center of communication, glamour and finance, is inhaling the debris of Cancer Alley.

As a result, part of New York itself—the borough of Staten Island, certainly, and probably parts of Nassau County, Brooklyn, and lower Manhattan—has become an extension of that pollution "alley." On Staten Island there is a monitoring program being set into place that will measure for compounds which had not been evaluated before. The state as a whole, as an example to the rest of America, is setting in place a network of regular toxic monitoring.

No wonder, either: The air is even a threat to the politicians. One of them, Guy V. Molinari, a conservative Republican who represents Staten Island, talks in a tone wherein fright is just barely masked by his bursts of frustration.

"I have a serious sinus problem," he says. "The first thing I do every morning is go to the bathroom and take a strong aspirin.

"My wife has an allergy as well. I have a child, a daughter, and she was sick from the day she was born and she's now twenty-eight, a city councilwoman, just elected last November to the city council of New York City. There were times we simply had to force her to school, sick as she was with an allergy problem.

"I lost a mother to emphysema and I lost a father to lung cancer. Neither one smoked a day in their lives. They abhorred smoking!"

Asked his reaction to that, he says, "Of course, there was sadness at first, and after you go through the period of grieving, you ask yourself the question, 'Why?'

"As I got into the problems of air pollution and realized they are as pervasive as they are, as deadly as they are, and how little is done by the

255

governmental agencies, well, then, yes, a sense of anger set in. Today I think I'm angrier than ever. A little over a month ago I came home and took my wife to supper. We got home and were alongside the condominium in which I live, walked out of the car, and my wife and I were almost knocked down by a new odor, one that we hadn't witnessed before—very strong, insecticide-type odor. It was so bad that we both ran for the door of the apartment house.

"It's hit every family on Staten Island if you've lived here long enough. I'm fifty-seven and since I was a child I can recall the odors. The odors may have been stronger then, but I don't think they were as hazardous to health.

"It's a continuous assault. We have patterns that we've studied over the years. We know, for example, that on Friday nights and Saturday nights between midnight and five in the morning there are those who abuse the atmosphere knowing that there's no monitoring going on during those hours. That's when they'll let their deadly booty go.

"I met a man—a couple—at a shopping mall last week and the man almost assaulted me. He was infuriated. He had lived on Staten Island for two years—a relatively young couple—and he said he loved the island but he said, 'How can you permit this to go on? My two daughters have been sick every day since we've been here. I think they're being poisoned, they're dying.' He didn't know [the things] I've been trying to do."

With what he believes to be "hundreds" of his constituents dying from the pollution each year, Congressman Molinari had pounded desks and heckled environmental authorities into coming out there and smelling for themselves. But the bureaucrats often arrived—even after the pleadings of a Republican congressman—hours after the toxic odors had dissipated.

"It's really an indictment of the system we have today. We're up against big bucks, let's face it. The industry has not shown me very much. We've caught them in lies, we've caught them releasing waste gases only when the prevailing wind was coming to Staten Island [avoiding the noses of New Jersey officialdom]. When you appeal to the New Jersey DEP, their prime concern is jobs—jobs, jobs, jobs at any price. The fact that it's poisoning and killing some of their own people as well as those who live downwind on Staten Island doesn't seem to get through to them. They're *blinded.*

"Bhopal was a dramatic event, the event that draws everybody's attention, [but] I think maybe more serious is that steady release of toxics on a daily basis. You don't have people knocked over and receiving emergency treatment. It's very slow, it's very deadly, but it's doing its job and you don't know it. The incidence of asthma is extraordinarily high. We're losing a lot of people and there's a lot of others who cannot function. One doctor told me there isn't a day in the week when he doesn't tell people to leave Staten Island."

Though by far the least crowded of New York boroughs, Staten Island, with a population that has reached 373,900, is not the isolated outpost it sometimes seemed. And its congested, wheezing people serve as a major representation of the 63 million Americans—about one fourth of the population—who've had some experience with chronic respiratory disease. (In fact, such illnesses are the most rapidly increasing of the ten leading causes of death, responsible for 12 percent of all mortalities).

Thus was it only apropos that in another burst of bluntness, a citizens group on the island named itself GAG.

When the winds come from the southwest, benzene seems to be a special problem, while toluene is heaviest from the north. The westerlies bring tetrachloroethylene, and you could just about always bank, anywhere, on some formaldehyde.

Two companies blamed most often for the island's episodes were Merck and Company of Rahway, New Jersey, which made worming pills for cattle and seemed to be the source of the cat-urine odors (though management denied this), and U.S. Metals Refining Company of Carteret, New Jersey, which shut down its smelter and certain other departments after New York State's aggressive attorney general, Robert Abrams, who had taken the lead in environmental matters among all state leaders, sued the New Jersey firm on claims that it sent lead and dioxin into the air.

It is a mere five hundred yards from the island, and according to a memorandum dated February 6, 1986, from the DEP's Division of Law in Trenton, there was particular concern about the firm (owned by AMAX) "considering the *combined* effects of lead and dioxin" (author's emphasis).

Mayor Edward I. Koch had also complained, his gripe the Du Pont and American Cyanamid plants in Linden.

When the wind shifts and comes from the east, New Jersey, in turn, is assailed by New York's abrasive particulates.

Whether or not it had anything to do with the air, there were cancer clusters cropping up in Brooklyn (the Flatbush area, where melanomas

257

were reported) and out on suburban Long Island. There were also the more blatant incidents: At least 370 people had been hurt by at least seven hundred chemical accidents in the state in a three-year period, and out in Farmingdale, fumes from burning pesticides had forced a thousand residents into evacuation, while a fire at a Queens electroplating plant around the same time threatened surrounding blocks with cyanide and injured eleven fire fighters.

Also in Queens, thirty-one people took ill from a pesticide which had been sprayed at a food-stamp center. There are also, in that borough, the steady area or "nonpoint" fumes from heavy highway traffic, petroleum storage, and both La Guardia and JFK airports.

Across the East River, in Manhattan, between Fifty-fifth and Fifty-eighth streets, a ruptured steam line spewed asbestos-contaminated debris over a three-block area.

The air in Manhattan, trapped in the concrete canyons, is comparable to—but not quite so bad as—the air in Houston and Los Angeles. There are the gas stoves in thousands of restaurants, the smoke from power facilities, the hospital incinerators, the spilled gasoline, the high-rise stacks. There are the fuming asphalt hoppers, smoky hot-dog vendors, dry cleaners with their chloroethylenes.

The collective issuance of tiny contributors, even the omnipresent can of roach spray, must count for something in the air. New York City has a population roughly equivalent to San Francisco, Chicago, and Los Angeles combined; each person adds his share.

Boiler smoke curls from one building to a neighboring skyscraper, dusting the very windows—about as direct as pollution can be. It takes only a few days for soot to gather on a piece of furniture and rob it of its gleam, or to smudge an eight-story apartment window.

In summer, on humid afternoons, the murk seems thick enough to chop.

Other times the skyline, far more immense than any other city's, is chimerical in a purple haze.

Monuments have deteriorated, and marble on at least one upstate building (in Schenectady) was found to be turning to gypsum crystals.

Using selective data such as that for benzene, one could argue that New York is not as intoxicated as Toronto, Stockholm, Zurich, The Hague, Sidney, or certain cities in England. It is most assuredly less polluted, by most barometers, than Tokyo or Mexico City.

While tetrachloroethylene, coming from all those drycleaning shops, is far higher than, say, Niagara Falls (and its trace metals—lead and zinc

—higher than St. Louis), its benzene is a fifth or even less than what some tests had shown for Los Angeles or Tuscaloosa.

On the other hand, the benzene in Central Park—where the joggers jog—is higher than Bayonne, New Jersey. Most of this comes from cars. There are 4.3 million gasoline-powered vehicles registered in New York City, Long Island, and the upstate suburbs of Rockland and Westchester. Some 16 percent of their pollution devices have been tampered with, and besides benzene, automobiles give off a mind-boggling array of other trace organics and metals, including one, toluene, that is pretty high throughout Manhattan. (It was detected at seventy-six hundred parts per trillion—higher than Newark—in that part of the Upper East Side, Yorkville, where I live.)

In Midtown, from Fifty-ninth Street to Thirty-fourth Street, and in other clogged parts of the city, such as Chinatown, Wall Street, and Greenwich Village, the pollution is worse. Were there bizarre health outbreaks and deformities here which, because of New York's size, had simply gone unnoticed?

Besides the dangerous benzo-(a)-pyrene, officials told me that if they look hard enough, they are almost sure to find some furans and dioxins.

As far as the outlook on volatile organics, one has only to know that mobile sources of them increased 10 percent over a recent five-year period. Every day more than 800,000 cars, trucks, and buses enter the twenty-two-square-mile island of Manhattan.

I am not worrying here about carbon monoxide. Each car is like a little moving chemical factory, emitting the combustive residues of synthetic fuel additives that, in the case of leaded gasoline, include those chlorinated ones which, upon heating, could form the most toxic furans and dioxins.

Among the highly mutagenic (and in one expert's memorable words, "horrendously carcinogenic") emissions are also nitro-PAHs such as nitrofluoranthene, which come from diesel combustion.

Nearly any car can give off a little cyanide.

"I've never looked for a compound hard enough and not found it," notes John E. Sigsby, a chemist at EPA's automobile testing grounds in Research Triangle Park. "The automobile or diesel engine is a very complex chemical reactor. It is capable of doing just about anything."

Before the automotive pollution devices were mandated, says Sigsby, the monoxide levels were such that the seemingly outlandish scenario of people actually falling over on urban street corners was not so outlandish. "And that," he says, "would have been embarrassing."

Nothing is more visibly unhealthful than New York's daily traffic of fifty-six hundred buses. Sometimes there are a dozen on a single block. They send toxic black puffs up to the rooftop gardens along Fifth Avenue. That they are diesel-powered is an unnerving fact.

Diesel particulates cry out for more study to see just how genotoxic and carcinogenic their toxics are. Their carbon cores may adsorb ten thousand different chemicals, and the diesel particulates are 30 to 70 percent more numerous than emissions from gasoline. According to the Natural Resources Defense Council (NRDC), some 6 million pounds of them are spewed in the city's air each year—at breathing level.

This output is expected to increase 11 percent by 1995, and one could also point to diesel-powered cabs and the tens of thousands of pavement-pounding trucks as other substantial sources.

For those millions breathing this exhaust, it is only fair to know that a study by the National Cancer Institute found truck drivers exposed to diesel pollution in Detroit to have 11.9 times the lower urinary tract cancer of an unexposed group. A similar study showed the same in London. Each year, said NRDC, from thirty to eight thousand excess lung cancers might be caused in America by diesel pollutants. The particulates are not only chemically overloaded but also of the size that can penetrate the lung.

It is small wonder that a Manhattan resident is five times more likely to be the victim of lung disease than he or she is to be a victim of homicide. For when the organic extracts of the particulates they breathe are painted on mice, they cause skin cancer.

In addition to all else, residents of New York City are especially vulnerable to toxic exposure *indoors*. Like the rest of Americans, they use and own hundreds of products that give off poisonous molecules. But, with 62,922 people per square mile, the living quarters in Manhattan are more closed-in and cramped. In other words, the air is slower, staler.

Indoor pollution is a type of contamination so considerable, so insidious, that it deserves sweeping legislation on the order of the Clean

Water Act. Stepping into an apartment building, a dweller might encounter chlorine, chloroform, or ammonia used to wash down a tiled vestibule or marble stairs; or, solvents from a door's varnish. The paint in an elevator (or anywhere else, for that matter) might be outgassing acrylic resins, xylene, or any number of similar volatile organic compounds. Down the hallway, trichloroethane might reach the lungs from textured carpet or wallpaper glue. Or, from old, exposed pipe insulation, there may be some fibers of asbestos.

Inside a typical apartment is a veritable taste of Louisiana. It is not unlike living near a toxic dump. One need not look much further than under the kitchen sink, in storage closets, or in bathroom cabinets to come up with a rush of polysyllables. A random survey of a typical bathroom might yield phenols in the skin cream, benzoate in the mouthwash, chlorhydrates in the foot powder, trichloroethylene in the mascara, and ammonium chloride in the gentle, silky shampoo.

Disinfectants might involve any number of biocides, including cancer-causing dichlorobenzene, which would also come from moth crystals and air "fresheners."

If there is a can of drain opener, that would mean caustic potash is on the premises, while sunburn spray—which one might try using upon a scalding by the potash—would emit a benzo compound.

In perfume, formaldehyde or trichloroethane in the aerosol cans. When tested by government scientists, one house had 350 detectable chemicals in the air. The average home contains forty-five aerosol cans.

Among cleaning agents are some of the most noxious compounds. The strong ones are frequently associated with strippers: paint, varnish, and tile strippers composed of such things as methylene chloride and amines.

Though, in the EPA study, xylene seemed especially prevalent, benzene, as usual, is also a prominent force—found in steel soap pads, liquid detergents, and furniture wax.

The troublesome formaldehyde, among many sources, comes from protective coatings on furniture, drapes, and carpets, or from cosmetics, household insulation, and particle board. Even fiberglass may be carcinogenic. The list is not endless but seems so most of the time: nitrogen dioxide from stoves, toluene from the print on fresh magazines, tetrachloroethylene on clothes back from the cleaner's, and other compounds volatizing from plastic or wafting in cigarette smoke.

Cigarette smoke contains not just nicotine but also such things as acrolein, aldehydes, nitrosamines, and aromatic hydrocarbons.

261

Perhaps as many as five thousand people are dying from ailments linked to passive smoking each year. "The most important source of benzene and styrene indoors," observes Lance Wallace, a scientist at EPA headquarters who worries about leukemia in children, "was cigarette smoke."

While chlordane had been banned in New York State, city dwellers are in a constant battle—chiefly a *chemical* battle—with rodents and insects. In the roach spray, which piles up its molecules in the small confines of an apartment, are pyrethrins, carbamates, and petroleum distillates.

To a certain degree, then, we all live in Midland, Michigan.

When it comes to pesticides, there is a different but also very significant risk out in the further reaches of the Boston–New York–Washington megalopolis. For many years silvex—carrying dioxin—had been a home weed killer. Golf courses and front lawns from Long Island to Massachusetts are treated with a galaxy of chemicals—the toxicity of which, too often, the applicators are ignorant.

Inside many New England homes, along with whiffs of lawn and tree pesticides, is probably the odorless and truly terrifying radon. The states of Connecticut, Rhode Island, New Hampshire, and Maine are suspected of having uranium-bearing soil that might release risky amounts of radon into the homes. According to Richard Guimond, director of the EPA's Radon Action Program, each year five thousand to twenty thousand lung cancers nationwide might be related to the problem.

The effect of combining radon with other indoor contaminants remains unknown. Again, "one" plus "one" might be more than "two." Toluene has been known, in combination with other chemicals, to slow down their leaving the body. Or, in the case of greeting a certain anthracene, of linking up with its cancer-causing effects. Exposed to ozone and nitrogen oxides, some constituents of particulate emissions become more mutagenic.

It can only be hoped that no New Englanders will have to go through what Stanley Watras in Pennsylvania experienced. He had to move his young family out of their home around Christmastime and rent an apartment for more than half a year, until basement cracks could be sealed and a special ventilation system installed, sucking radon out from under the house. "People were saying I was going to die and had my kids dying too. My house is like a uranium mine."

But what is transgressing inside a house cannot preclude what is

262

happening outside. On a beach near New Haven, just across from Long Island, I watched a brownish-red smog coming in from New York City like a tide of burnished dirt.

What "haven"? From Richmond, Virginia, to Portland, Maine, ozone—that herald of hydrocarbons—is stubbornly resisting efforts to bring it into compliance with federal regulations.

"Washington, D.C., puts out its organics, its nitrogen oxides, and by the time we get to Baltimore, it affects people there," noted William S. Baker, chief of the EPA's air programs in New York. "Baltimore adds its own contribution, and it moves up to Wilmington and Philadelphia, through New Jersey, Connecticut, and Boston."

The pollution from Boston is capable of going out to sea, then washing back when the wind shifts and landing in Maine and Canada.

On Cape Cod were reports—perhaps a water problem—of towns with cancer mysteries.

The Northeast corridor is home to 31 million people, and that kind of density, said the EPA's administrator, Lee M. Thomas, means there would be major quantities of toxics in the air even without the heavy industry. He recalled the agency's report that claimed (in questionable fashion, of course) that almost 75 percent of cancers might be due to air toxics that come from nonpoint sources. "These are small, individual sources, such as drycleaning plants, degreasing operations, cars and trucks, gasoline sales, and woodstoves and other small units for burning fossil fuels."

Like Colorado, there are woodstoves and crackling fireplaces throughout the ski country of New England.

Indeed: the land of sugar maples. Except that throughout the region sugar maples are contracting sickly yellow spots. And the ashes have developed purple splotches. And the leaves on dogwoods are curling skyward, as if begging it to stop.

In Greenwich, in Stamford, and finally out in Bridgeport, Connecticut (where old foundries, brick warehouses, bulky harbor smokestacks, and one firm that *manufactures* furnaces, are all a mighty presence), ozone levels often have been as high or higher than in New York City—accepting as they do both New York's and New Jersey's ozone precursors, which take a while downstream to cook up.

By one ranking, urban Connecticut was the third-highest ozone region in the nation, below only Los Angeles and Houston.

After years of decline, the early 1980s saw ozone exceedances shoot back up. "It's a ball-park figure, but there are probably two million people at risk in southern New England because they are exposed to unhealthful levels of ozone," says EPA's Marvin Rosenstein.

Vermont, said one official there, receives pollution chiefly from eight states, including Tennessee and West Virginia. About 10 percent of the trajectories over New York find themselves over Connecticut three hours later, according to another study. Emissions from cars and factories in Boston cause more ozone over Nashua, New Hampshire, meanwhile, than over Boston itself.

Given the region's level of urbanization, and the fact that areas such as Boston were historical starting points for industry and thus for the hazardous materials that go along with it, it should probably not be surprising that cancer breaks out frequently. But there is still something riddling about the fact that Massachusetts, Rhode Island, Connecticut, and New Hampshire were all in the top ten of 1970–1979 cancer states.

Why was Rhode Island higher than New Jersey?

Yes, there had been the smoldering toxic dump on a pig farm near Providence, and there are jewel makers using solvents, electroplaters with their various compounds. Dichloroethane, toluene, and phenols were detected near one chemical plant a dozen or so miles south of Providence, and when Congressman Molinari of Staten Island had one of his staff members take a quick survey of the national situation, the subsequent report, "Ill Winds," said there were 122 industrial facilities in the Providence area.

But it is not as densely industrialized as a good many other areas, and so one wondered: Was something *undetected* at work?

In Maine and Vermont, paper, printing, varnishing, leather processing, and huge dry cleaners were responsible for the hydrocarbons detected in such places as a nearby school. Just east, in rustic New Hampshire, the key concern was what volatized off the state's dumps. Near waste sites or above polluted brook water in Salem and Nashua, such compounds as xylene or hydrofurans were logged in the air.

Above, from the troposphere, America's poisonous residues—having caused many more deaths, it would appear, than the United States government could reckon—continued, indeed, to fall invisibly and odorlessly upon the land, invading our youth.

Or the cloud headed across the Atlantic to join the aerosol from other nations and other continents, in the swirling global mix.

In Reading, Massachusetts, a woman named Judy Broderick had searched through old death certificates and counted too many brain tumors.

Not unlike Houston and Chicago, Boston itself has xylene and toluene at up to four thousand parts per trillion.

Between Bridgeport and Hartford, residents near an American Cyanamid plant in Wallingford felt they were being affected either by the plant or by sewer gas. They said the pollution even caused hallucinations. "My oldest daughter, Jennifer, started to cough deeply," said one resident, Janice Nuzzo, of an incident a couple years before. "The family dog lying at her feet suddenly jumped and clawed her face. She ran into the bathroom for a cloth and came out screaming that her eyes were burning.

"I called my physician, who informed me he had five people in the office that day with the same complaint. He ordered medication and we hung up. Immediately, my daughter broke into large blotches and welts all over her neck and chest. About an hour later, the glands in my neck, underarms, and groin swelled and the right side of my head went numb.

"This sensation lasted about an hour. That evening, Jennifer suffered acute abdominal pains and vomiting. My middle child, Aimee, tossed and turned and my toddler cried all night. Experiences like these can be recounted by numerous families in Wallingford. It is becoming increasingly apparent that we, the residents of Wallingford, Connecticut, are living under a toxic cloud which centers over us but is shared with neighboring towns."

In a letter to the state's environmental commissioner, she said, "For the love of God, send us help."

In Connecticut as well as Boston and the rest of New England, the focus of concern is beginning to revolve around garbage incinerators. Put some plastic or other chlorinated and brominated materials into the refuse—all that plastic wrapping! all those plastic bottles!—and dioxin of the TCDD type, in some quantity, is bound to form.

Officials in Massachusetts ordered dioxin tests on the ash from trash-to-energy plants in Saugus, North Andover, and Pittsfield. By 1987, wherever it was looked for, it had been found.

Up to fifty such plants will be built in the Northeast by 1990, including new burners in New York City, on Long Island (where state law requires landfills to close very soon due to water pollution), and in Westchester, where the county's last landfill is already closed. Themselves concerned about the emissions, officials from New York, Connecticut, and Rhode Island have had to ask the federal government to adopt new rules governing municipal incinerators.

The concern of the average person, with good reason, is that government estimates of dioxin (and furan) dangers have been based upon guesswork and are understated. Critics looking at incinerators in such places as Peekskill, New York, declared state assessments of the cancer risk to be between eight and seventeen times too low. Others underscored a growing suspicion that the dioxin fallout accumulating in nearby cows (and thus the milk) is a greater threat—perhaps a far greater threat—than direct inhalation.

Not that a bit more dioxin is needed. In a paper presented at the annual meeting of the American Association for the Advancement of Science, two researchers, Thomas Webster and Karen Shapiro, and a very well known ecologist, Barry Commoner, all from Queens College, argued that, contrary to federal assessments, the lifetime cancer risk from chlorinated dioxins and furans is already 330 to 1,400 per million people, higher than the risk for benzene and that, if correct, means that exposure to all forms of furans and dioxins might cost 200,000 people or more a life-or-death struggle with a malignancy.

As the 1980s give way to the 1990s, the burning of chemical-yielding garbage promises to help thrust toxic air pollution into the limelight it so richly deserves, and to open up a new era of environmental concern. It is an issue within an issue but one that can be expected soon to challenge the predominance of landfills on the environmental agenda.

For now, however, the dangers from indoor pollution, from diesel motors and gasoline stations and a few million other little businesses, from huge refineries, from the twelve thousand chemical manufacturing plants, from the many units *within* each large factory, from chemical ponds and storage tanks, make the issue of toxic air pollution an extremely crucial and major one with or without the swelling issue of trash incineration.

And no government at any level truly knows—or wants to know—the full reality of this.

In that report from Congressman Molinari's office, "Ill Winds" (a decidedly "unscientific," yet brave, challenging study conducted by legislative assistant Michael Torrusio), 156 counties in twenty-seven states were examined for respiratory cancer rates: "The study found that those counties that have significantly high death rates from respiratory cancer either have, or are downwind from, a petrochemical complex.

"Perhaps the most astonishing observation to come from this study is that despite the total lack of petrochemical industries, the rates of respiratory cancer are high in those areas where the three air systems converge. Northern New York State, Vermont, New Hampshire and Maine all lie within the north leg of the confluence of the three air systems, and within these states, those counties exposed to this air flow have high cancer rates although none have any petrochemical industries."

Protesting the 400 million or so pounds of hydrocarbons emitted from New Jersey, a group of Molinari's Staten Island constituents assembled at the Goethals Bridge toll plaza and held placards in a blistering wind. The protesters also shouted slogans in the din of truck traffic, cheered on by some passing motorists, cursed by one or two others.

Most of the passing looks, however, were looks of puzzlement, for Americans, even in the most experienced cities, simply did not understand—as their government scientists did not yet understand—the extent or implications of airborne toxicants.

"The situation is getting worse, and the people have run out of patience. They're becoming violent," Congressman Molinari fretted to me.

Soon after, however, the citizen protests seemed to have caused a very marked reduction in the New Jersey odors. It was a testimony to how groups of average citizens across the country had started to make a big difference. In Chicago and even Louisiana, where the crisis raged out of control, citizens, once they awoke to what threatened their children, had made their voices heard, and little victories were won as a result.

Though the odors had lessened, one of the Staten Island leaders, a fireman named Joe Procopio, feared those toxics that carry no odor and are still around. "It's another Love Canal," he said. "But Love Canal was in the ground. This is in the air."

PART VI

SUNRISE AT THE LOVE CANAL

26

Where once there had been rows of homes, now there was a huge burial mound.

Where once there had been a school there now were pipes to decontaminate the rainwater.

Where once there had been public streets, now there was a tall green fence to keep out the curious.

Where once there were shrieks from the playground, there was only the caw of a lonely gull, and the complaint of birds on a lifeless tree.

About seven hundred houses and a large federal housing project were evacuated when health problems were found in the area, and on the outer periphery of the Love Canal—spookier than Times Beach—there were basketball hoops that would never again swish, and swimming pools that never again would feel a youngster's excited waves.

Where once there were neighbors waving from their porches, now there was no motion except for the rustling wind.

The neighbors I had once known were all gone now, scattered everywhere, trying to start new lives and forget the terror.

In some homes black sludge had oozed through basement walls, or flooded the backyard.

An emergency was declared, and the people left hysterically.

271

Now, silently, a few miles away, a smokestack purpled the sky as its plume refracted across the horizon.

In the back of my mind, in the gloaming, I could still hear those Love Canal kids: *Mommy, let's get out of here before the chemicals kill us! It's gonna get us all! Nobody's safe!*

Or the man in a packed, steamy auditorium who had screamed at the bureaucrats, *You're letting us die, dammit! You're going to stand there and watch us all die!*

Behind the evacuation zone was the contaminated creek near the home of a little boy who had played in the muck and who had died suspiciously.

I could still see his mother back there, staring hard at the water, remembering her son's final days . . .

At the other side I could hear the sobs of a woman who'd just had another miscarriage, and near her boarded-up house, the rattle of a loose warning sign.

A warning sign could be hung somewhere in every city: *Danger, Toxic Air Contamination.*

This time there will be no evacuation. This time there is nowhere to hide.

The poisons once thought to be a serious concern only near a place such as Love Canal are known now to be everywhere. They appear in the air of an Alpine forest, or over a Pacific island. They are also in the cabinet under the kitchen sink.

Solvents, refrigerants, methane, and oxides—all are among the pollutants that are increasingly gathering in the atmosphere. There they may be dangerously altering the natural ozone layer that protects us against harmful radiation from space, or are causing the "greenhouse" effect, blocking the escape of heat from the earth's surface and, in so doing, threatening to raise the planet's temperature to the point where lush farmland would turn to dust bowls, and polar ice would melt.

There is now a "hole" in the protective ozone above Antarctica that is about the size of the United States.

Fed by the melting ice, oceans may one day rise to inundate major coastal cities, while epidemics of skin cancer, spawned by the ultraviolet rays, would spread around the globe. It is a mega-problem caused not just by America but more so by the combined emissions of Japan, Eu-

272

rope, and Russia, "a totally uncontrolled experiment," in one expert's words, "with no kind of knowledge of where we are going in the end."

We are doomed only if we bury our heads in the sand. The Love Canal proved that Americans could be marshaled into decisive action. It led to a nationwide cleanup effort, the "Superfund," that will attend to hundreds of similar dumps. And there also have been victories in controlling what goes into the air. While toxic pollution remains uncontrolled, the United States has come a long way in decreasing the more conventional contaminants. Since 1970 carbon monoxide has been reduced by 34 percent and total suspended particulates by more than half.

Now, in the same way that we learned about ground pollution from the Love Canal, or smog episodes from disasters in London, we can learn about exotic atmospheric molecules from the Kanawha Valleys and Houstons and Midlands.

They constitute, like the Love Canal, a full-fledged synthetic plague: When radon and other indoor pollutants are tallied alongside petrochemical emissions, pesticides, wastewater evaporation, airborne metals, and combustion fumes, the death toll is perhaps as great or even greater from toxic air pollution—from "the toxic cloud"—than from a disease like AIDS.

We live in an era during which life expectancy as a whole has increased because of better health care for conventional types of physical distress. But we are also at the point where cancer is contracted by 30 percent of the population. Of Americans now living, 74 million will be hit with it. The cancer death rate has increased 26 percent in just two decades.

There are also the respiratory ailments which bedevil America's youth as no other class of disorder does, and the seemingly minor ailments, like ear infection, that crop up often and might point to profound disturbances in the immune system.

Birth defects are on the rise, and we must find out why. Between 1970 and 1980 there was a 300 percent increase in reported cases of displaced hips and a 240 percent increase in babies with ventricular septal defects—a hole between chambers of the heart. There also has

273

been an increase (though not statistically significant) in infants born with shortened limbs. About 12 percent of the babies delivered today will be born with a significant health problem.

At least part of the reason for such increases is better diagnostic and reporting techniques. Still, the question remains: Are we watching the nation's genes being dismantled in slow motion?

Is the placenta now a pipeline of poison to the fetus?

In addition to asthma, emphysema, and nerve disorders, there has been speculation that environmental pollution might also figure into puzzling diseases like Alzheimer's. Research reported by Canadian scientists suggests people living near pulp mills and petrochemical facilities have a strikingly high chance of contracting Parkinson's disease, a problem that afflicts perhaps one percent of the world's population (or 50 million people).

Some scientists have estimated that only 1 to 5 percent of cancers are due to pollution in the surrounding environment. Some guess much higher. At 5 percent, however, this would mean 24,150 people dying in the United States each year.

"That's an estimate that's often repeated," says Dr. Irving Selikoff of Mount Sinai Hospital in New York. "But there are several problems with that. First of all we don't know what causes most cancer. And so when you're saying five percent of all cancer, it implies you know the other 95 percent."

"Actually, we know the cause of about 35 percent of cancer. So if it's 5 percent, of the 35 percent known, it would really be *one seventh* [or about 14 percent] environmental."

Dr. Selikoff describes studies such as the EPA's "Six-Month Study" (estimating less than two thousand cancer deaths a year from air toxics) as being "worth the paper it's written on. . . . No one is exposed to only one agent in our environment. And they add up. We don't know what the agents are, we don't know what the multiple effects are, we haven't measured for cumulative effects, so we're sailing on a sea of ignorance and the port that we reach, the shores that we reach, are 450,000 cancer deaths a year."

The potentially cancer-causing or in other ways damaging agents in our environment increase with each passing week. Soon they may include new products from genetic engineering and residues from chemical weapons that the Army wants to incinerate.

Instead of urgently grappling with the problem, our government, at all levels, has avoided the issue and in some cases has made it worse. Recently, as just one example, the FDA proposed rule changes that in effect will encourage wider use of the plastic polyvinyl chloride for packaging beverages and food.

The more packaging, the more plastic in the garbage. The more plastic in the garbage, the more chlorine. The more chlorine, the more furans and dioxins from our incinerators.

The production of plastic resin has grown at an annual rate of 12.7 percent since 1972. Alongside solvents and pesticides, plastics constitute the most serious cause of highly toxic pollution. Yet because they are an easy way out—a convenience we suddenly can't do without—we have allowed their use to expand rampantly. Plastics are replacing our less hazardous products such as steel, glass, and wood, tossing nonbiodegradable components into the fragile ecosystem.

As was initially the case at the Love Canal, the experts don't yet view this as any sort of national emergency. "I guess I would have to characterize the issue as one in which there are problems in some areas," the EPA's administrator, Lee M. Thomas, told me, adding that as far as immediacy, toxic air is more of a problem than toxic groundwater.

"In other words, I think there are areas that have fairly high emissions and in the past people have called them 'hot spots' or whatever.

"The kind of hot-spot issue I'm talking about is places where you've got not only 'area' sources but a cluster of point sources, like the Kanawha Valley or places like that. So I couldn't give you a generalization of the issue, because it has a lot of dimensions.

"Beyond that, I think there are general problems that can't be addressed. I don't think they're of a magnitude to cause a tremendous amount of concern, but I think they're of a magnitude that requires action. And I'm thinking more in terms of area sources than I am of point sources. For instance, I think there are problems with woodstoves and I think they need to be addressed."

Woodstoves?
No one could doubt that they are a very substantial concern. But it

seemed as if the EPA, under the Reagan doctrine, was intent on concerning itself mainly with those problems which could be laid to private individuals or small companies and not big businesses.

Either that or it merely sloughed thorny issues off. The agency already had shifted responsibility for controlling emissions of chemical factories to the individual states where they're located.

While the EPA had recovered under William D. Ruckelshaus and now Thomas after the early debacle of Anne Gorsuch-Burford (to where it was about as effective as the EPA of Jimmy Carter and in some ways probably offered an improvement in professionalism, with congratulations in order for the agency's dispensing of some record-breaking corporate fines), too many of its fourteen thousand employees were badly underutilized and in no hurry whatsoever to uncover new problems; far too many poisons remained unregulated; and the President himself seemed completely unaware of the mounting toll from toxic pollution, which held thousands hostage and terrorized right here at home.

The President wanted to believe that pollution was largely a problem of the past. During his first campaign he had once even said something to the effect that trees are more of a source of pollution than man is (they *do* give off hydrocarbons), and he ignored the EPA so blithely that at one point he forgot the administrator's name.

There was also the time President Reagan appointed industrialist Armand Hammer to head a special cancer panel. It is difficult to describe this appointment with anything but disdain, for at the same time Hammer was supposedly leading efforts to cure the disease, he was also chairman of Occidental Petroleum, which owns the very company that dumped the chemicals—the carcinogens—into the Love Canal.

The nadir of Reagan's environmental policies may have come during the reign of Burford, but a close runner-up was the President's veto, in 1987, of a $20 billion bill to clean up the nation's waters. After years of improvements in sewage disposal precisely because of similar legislation, the President wanted to set back the clock.

But in a major, well-planted rebuff, the House voted 401 to 26 and the Senate 86 to 14 to override Reagan, showing just how out of touch he was with the American people's environmental sensibilities.

However, many problems at EPA were and are problems that predated President Reagan. It is a wasteful bureaucracy that has been this way for more than a decade, overlapping, convoluted, and woefully shortsighted, granting insufficient concern for the trace contaminants which will plague future generations.

In fact the EPA's bureaucracy is as inefficient and confusing as the headquarters itself. The building is known as Waterside Mall, and it originally had been built as an apartment and shopping complex before the government, treating the agency like some kind of orphan, housed it there.

The offices are impersonal and numbered in a fashion that even longtime employees have trouble figuring out. One can spend half a morning lost in an isolated wing which seems disconnected from anywhere else.

As if to symbolize the disarray, there is one part of headquarters where a person must go *up* a set of stairs in order to find an elevator that will take him *down*.

Cubicles have a disheveled, transient look, with unread documents slouching off the bookshelves, and wires dangling from the ceiling.

At the desks, it is sad to say, too many workers have become paper-pushing robotons with no deeply expressed affinity for the environment.

They are hidden in an agency which is broken into various "offices" or "laboratories" which are broken down in their turn into "divisions" which are broken down into "branches" and "sections" and "teams." There might also be an "office" *within* an "office."

The result is wasteful duplication, or data that is scattered between offices in such a way that a person dealing with a certain matter in one outfit is not aware of a pertinent study that came out of another outfit.

And the end results of all *this* are the rasping lungs and birth defects that neither bureaucrats nor the President ever see.

So the scandal is the insidious kind, the kind that comes from negligence and lack of knowledge, from compromised policies that subtly transform the government employee into industrial lobbyist.

EPA, at the bottom line, has been too lenient with industry.

Indeed, sometimes it is difficult to tell if one is speaking to a federal official or a corporate salesman. Officials at the Louisiana attorney general's office told me they had taped a conversation with one EPA official who was trying to stop their opposition to a proposal by Rollins to burn PCBs in its problem-plagued incinerator. The official seemed to hint that if the state let Rollins have its way, Louisiana might be chosen as the site for a new naval base.

Always more accommodating to corporations than individuals, no matter *who* is president, the EPA was in no hurry to expand its regulation of hazardous substances. In the seventeen years since enactment of the Clean Air Act—which, under Section 112, calls for control of toxic substances that might threaten human health—the agency had categorized only seven substances (arsenic, beryllium, mercury, benzene, vinyl chloride, asbestos, and radionuclides) as officially "hazardous" and thus subject to regulation.

At one point in 1984, by which time only 5 of 650 targeted chemicals were being regulated under Section 112, Congressman Gerry E. Sikorski of Minnesota commented, "At that rate it will take 1,820 years to do something about the remaining chemicals on the list, almost as long as since when Christ walked on Earth."

By March 1, 1987, the EPA had decided *not* to regulate such common and troublesome pollutants as phenols, acrylonitrile, toluene, chlorobenzene, methyl chloroform, and benzo-(a)-pyrene (though coke emissions have been placed on an "intend-to-regulate" list).

If a compound was on this official list, it meant EPA—slowly and not so surely—was developing regulations for the chemical. If it was not on the intend-to-regulate list but instead the "intend-to-*list*" list, that meant the agency was deciding whether it should be on a list of compounds set for final regulatory calculations. In other words, it was a list of compounds about which the agency would decide whether to decide!

A mere twelve chemicals were being assessed in detail (including hydrogen sulfide and phosgene) while twenty-four others were under only preliminary screening (including styrene and xylene). It was hoped that by 1988 eleven more compounds could be added to the seven that were being regulated—an important but still a dishearteningly small number—and the agency was especially slow coming to final determinations on two widespread and dangerous metals: chromium and cadmium.

If all this was not discouraging enough, the EPA had gone back to the lonely list of seven regulated compounds and had modified strictures so that one of them, vinyl chloride, could be emitted in certain amounts where before no preventable releases had been allowed.

This could not have been welcome news to the 4.6 million people living within five miles of a vinyl chloride plant. But, then, EPA did not have much of a handle on who lived near what. "In fact," complained a congressional report, "the EPA was unable to provide an up-to-date list of where the plants in this country are located."

Also, industry had created brand-new compounds which it had not reported to the EPA.

The first systematic survey of routine toxic releases was initiated not by EPA but by the House Subcommittee on Health and the Environment, led by Congressman Henry Waxman of California. It was a very rough, preliminary study of eighty-six large chemical companies and it relied almost solely on information volunteered by the companies.

Still, the subcommittee found 196 unregulated chemicals—including the MIC—being regularly released into the nation's air. "Almost every chemical plant we received information about is releasing staggeringly high rates of hazardous chemicals, even in routine releases," said the congressman.

What regulations we do have are based, too often, on flimsy data. More and more, federal and state agencies are coming to rely upon formulas of risk assessment to make decisions on new laws and other protective actions.

In its simple form, such a formula estimates the level of exposure to a given populace and then, using the experimental results from animal studies, estimates how many deaths the exposure might cause among human beings.

The estimated effect on humans is essentially weighed against how much it would cost to remedy a particular problem. The kind of questions being asked, in plain language, place a price tag on human life. For example: If exposure to a chemical will cause two extra deaths, are those two people worth the $1 million, or $2 million, or $12 million it might cost to fix the problem and stop the exposure?

While supporters of this coolheaded (and coldhearted) method of evaluation argue that it usually overstates the risk (and therefore, when it errs, does so on the side of safety), others just as effectively reply that risk assessment is reheated hoodoo: assumptions built upon other assumptions, sort of a mathematical way of expressing intuition.

A key problem is that risk assessment assumes there are thresholds below which the chemical in question will cause no deleterious effects, when, to the contrary, we have *no* definitive idea what the effect will be of steady, low-level exposure over a long period of time.

Estimates often range wildly. The calculations on how many lives

279

may be lost can differ by a millionfold or more. "By varying the method of low dose extrapolation used, and the toxicology or epidemiology study which formed the basis of the risk assessment, commenters to the OSHA policy developed risk estimates for exposure to one part per million of vinyl chloride which ranged from *one in 100 million* to *one in ten,*" admitted an EPA brochure on the topic. "A similar exercise with saccharin . . . estimated the expected number of cancer cases in the general population at *between .001 cases* per million exposed and *5,200 cases* per million exposed" (author's emphasis).

Nor can we rest assured that if an outbreak erupts, the medical community will immediately notice it. A good definition of a catastrophe, someone once remarked, is an outbreak so obvious that even an epidemiologist notices it.

The same insensitivity is true of animal studies. Assume, wrote one critic, Dr. Samuel Epstein, that a "particular agent carries a risk of producing cancer in one of 10,000 humans exposed; this would result in approximately 20,000 cancers in the United States population.

"Then the chances of detecting this in groups of 50 rats or mice, tested at ambient human exposure levels, are very low. Indeed, samples of 10,000 rats or mice would be required to yield one cancer, over and above any spontaneous occurrences; for statistical significance, perhaps 30,000 rodents would be needed."

It should be remembered that with the drug thalidomide, which caused gruesome birth defects in humans, it took a hundred times as much to cause such problems in rats and perhaps seven hundred times as much to affect laboratory hamsters. Other chemicals of the aromatic amine class are carcinogenic in man but not in guinea pigs.

The risk models do not—and cannot—factor in the nearly endless possible effects that chemicals might cause in combination with each other: the wild card of synergism.

When managers at a workplace in Germany replaced some of its n-hexane with less toxic methylethylketone because the hexane had caused neurological problems elsewhere, there was a sudden explosion of nerve toxicity at the plant. The "safe" ketone compound, it so happens, was potentiating the toxicity of the remaining hexane. In other cases the body's metabolism can turn a relatively innocuous chemical into a hazardous one.

The federal government is of course in no hurry to learn about such effects. A panel of scientists has charged that the Office of Management and Budget is seven times more likely to reject studies with an environmental or occupational health focus than to reject those concerning infectious or other conventional diseases.

Finally, there is the vital issue of ambient air monitoring. Aside from spot sampling or special studies, neither the federal nor state governments have a credible, constant, and comprehensive system in place that would detect a sudden, invisible, and perhaps instant onslaught of highly potent (and of course unregulated) substances like the furans and dioxins.

Instead we depend chiefly upon a routine of inspection whereby smokestacks are periodically climbed and tested for their gas and soot, records are reviewed, or the plume is visually checked to see how black it is.

Even these rudimentary safeguards are all too often random and poorly conducted. During one recent year, according to the General Accounting Office, about 40 percent of the factory inspections carried out by state and local inspectors were in some way inadequate.

Nor, in detecting the early warning signals of an exposure, are doctors much help. Rarely do they link an ailment with a chemical effect, for doing so would not only be "unscientific" but would also upset the industries who keep referral lists of local physicians for their workers. Put plainly, a local doctor stands to lose patients who work at the factory. Blaming a chemical might also ruffle feathers at the country club or otherwise hurt a physician's social climb.

Mostly, though, the problem is that toxic chemicals were not taught in their medical texts. They have been caught off guard, with no frame of reference.

"Physicians who were trained and grew up in a totally different world now find themselves unable to answer the questions of their patients, unable to answer the questions they themselves have," noted Dr. Selikoff, the researcher widely credited with defining the asbestos danger.

Industry shamelessly takes advantage of our scientific uncertainties and keeps pushing out new, more complex molecules, arguing, in effect, that chemicals should be treated like people: again innocent until proven guilty. They go so far as to downplay the significance of furans.

When I chatted with Harold J. Corbett, a senior vice-president of Monsanto in charge of environment, safety, and health, he maintained

281

that there was not even enough epidemiological data to indict his company's notorious PCBs. "What bothers me is that we have a lot of individuals who are promoting the fear of chemicals for their own purpose."

Man-made carcinogens are a "non-issue," he said, since there are "thousands" of natural carcinogens as bad as what industry makes. He himself has eight children, and if he worries about anything, it's not possible exposure to furans but instead their smoking and overeating. "There are essentially no threats to them from synthetic [chemicals]."

Were there the money and will to conduct the major kind of investigations necessary to dispel doubts about chemical effects, the next problem would be in finding an unaffected comparison group—the "control"—to measure against a group that has incurred exposure. The reason: Everyone in the United States, and in most developed nations, has now been exposed to compounds such as dioxin. In America we all have about 7.2 parts per trillion of the TCDD isomer in our fat. And so, it is difficult to separate out dioxin-related ailments from "normal" illness precisely because the country is now so pervaded by chlorine.

Nonetheless, there remains the chilling possibility that massive, across-the-board, but nearly undetectable, effects already may be taking place throughout the population—ever-so-subtle damage, perhaps, to our immune systems.

We are in a society which can claim to have toxicological information about just 20 percent of the chemicals it uses. Ours is a stunningly apathetic society which watched with little complaint as production of synthetic organic chemicals swelled from less than 10 billion pounds before World War II to 350 billion pounds by 1980—and still is climbing.

Ours is a society that cares little about the 2.2 billion pounds of toluene released into the air each year.

Ours is a society that disdains recycling.

We are also a society which will increasingly encounter furans and dioxins as the nation's landfills close down and incineration of wastes becomes, by default, the popular alternative.

While incineration may be better in some ways than dumping untreated poisons into the earth, it should never be considered anything but a temporary and highly risky solution to the problem. When potent compounds are involved, the burning of them should be under extraordinarily well-monitored conditions in isolated desert and prairie, or hundreds of miles out at sea.

Even this is not acceptable as a permanent solution, however, for incineration ships may leak or sink. The chemicals may be unleashed during transport to the ports. The smoke may rain long-lived toxicants into the vulnerable sea.

While we search for truly profound ways to halt exposure, certain other measures must be pursued without further delay. In addition to setting up a nationwide monitoring system, the regulators must greatly expand the list of restricted air pollutants and increase the fines for unleashing them—including not just higher economic sanctions but the implementation of prison terms for the executives responsible, since harming humans with chemicals is a form of assault and manslaughter.

As is the case with the common thug, such harm comes from unchecked greed.

Technically speaking, there are hundreds of little actions and pieces of equipment that can reduce the atmosphere's toxic load. Insulation, pressure blanketing, and floating roofs can be used on storage tanks to minimize fume buildup and release, vapors can be burned off, better precipitators can precipitate more particulates, better scrubbers scrub more dirt, or tighter filters filter them out. Vents can be improved with condensers to treat the vapors, and nozzles can be fitted on gasoline pumps to prevent hydrocarbons from escaping.

The use of coal and diesel fuel must be far better controlled, and activated carbon should be more widely employed to filter out the troublesome organics, while waste piles should be covered with special liners.

We must also focus more attention on the fine particulates.

Benzene emissions from chemical plants can be reduced by as much as 99 percent with less than 3.5 percent increase in the price of chemicals made with it, the American Lung Association reports. In fact, tighter environmental controls can be a *boost* to the economy. In addition to saving medical costs, stricter regulations will bring out an enterprising spirit. In 1985, spending on pollution-control equipment created nearly 167,000 jobs.

It has also been found that the tightening up of an incinerator to limit air from leaking can reduce dioxin emissions from eighty nanograms per cubic meter to a mere six.

As for the compounds once they are in the body, much more research is needed to find means of expelling them. Already there have been indications that certain drugs—or "chelating" compounds—can carry stubborn metals or chlorinated compounds to excretion, and a regimen of vegetable oil, vitamins (especially C, A, and E), plus exercise

283

and a proper, enzyme-building, intestine-cleansing diet also probably help.

In the end, however, the solution must be pronounced at a philosophical level. And it is a seemingly drastic one. In the end highly toxic chemicals must be granted a wary respect—and isolated handling—not unlike what is practiced with radioactive elements.

Since industry has shown itself to be indifferent to the public risks it creates—haughty of heart and presaging a calamitous fall—we must move toward a policy by which no company will be allowed to make or use any compound until the firm proves that it can either disassemble that compound into harmless, natural compounds—destroy it completely, with no toxic residues—or proves that the compound has absolutely no health repercussions whatsoever.

Those compounds that fail the test but already exist must be gradually phased out if not outright withdrawn.

Industry must now assume the burden of proof. We can no longer endure its wanton poisons, nor can we any longer look upon science as a benevolent, all-powerful goddess. Too often scientists create more problems, and prompt more riddles, than they can solve.

Our society has two basic choices: to barter its health for modern conveniences and stop worrying about this all; or to begin ridding the environment of compounds that threaten cancer, that threaten immune disorders, that threaten the very planet itself because of their potential to ruin protective layers of the upper atmosphere.

In other words, we must prepare to ban a great many more chemicals than have so far been banned. And we must wean ourselves off certain conveniences—especially plastic—that cannot be made from biodegradable products.

Our laziness and selfishness, our uncaring, threaten to unravel us all.

So we are at Love Canal again, in the sense of standing before a new and compelling problem, in the sense of staring once more into the vast dark. But toxic air pollution should be viewed as a challenging and solvable problem, and therefore not so much a terrifying problem as an energizing, galvanizing one. Like Love Canal, the problem of air toxics has been buried for too long.

In the twilight, at the canal itself, are other memories as well. One tries nearly to squelch the recollection of the little blond girl who screamed at the top of her lungs as they poked a needle in her arm, looking for blood, looking for chemicals.

And the other little kids who had become so obsessed with the danger that they had taken to drawing tombstones with their parents' names on them.

Now, in their own twilight, the same thing was beginning to happen in New Jersey, California, Michigan, and Louisiana.

Would they shed light on this problem as Love Canal had on toxic dumps?

Indeed, twilight: The word has two definitions. One is the light diffused by the reflection of the sun's rays from the atmosphere just after a sunset.

But it can also mean the light that comes just before the sun begins to rise.

ACKNOWLEDGMENTS

My deep thanks to the hundreds of people in virtually every state (and province) who added something to this work.

Thanks also to my family for the strong support; to my agents; and to Craig D. Nelson at Harper & Row, whose input was invaluable.

INDEX